Youth in the Digital Age

Young people spend a significant amount of time with technology, particularly digital and social media. How do they experience and cope with the many influences of digital media in their lives? What are the main challenges and opportunities they navigate in living online?

Youth in the Digital Age provides answers from a decidedly interdisciplinary perspective, beginning in a framework steeped in context; biography; and societal influences on young people, who now make up 25% of the earth's population. Placing these perspectives alongside those of current scholars and commentators to help analyse what young people are up against in navigating the digital age, the volume also draws on data from a five-year research project (*Digital Media and Young Lives*). Topics explored include well-being, privacy, control, surveillance, digital capital, and social relationships.

Based on unique and emergent research from Canada, Scotland, and Australia, *Youth in the Digital Age* will appeal to post-secondary educators and scholars interested in fields such as youth studies, education, media studies, mental health, and technology.

Dr Kate C. Tilleczek is Canada Research Chair in *Young Lives, Education and Global Good*, and Full Professor of Education at York University, Canada.

Valerie M. Campbell, PhD (c), is a sessional lecturer in Sociology and Anthropology at the University of Prince Edward Island, Canada.

Youth, Young Adulthood and Society

Tracy Shildrick, Newcastle University, UK
John Goodwin, University of Leicester, UK
Henrietta O'Connor, University of Leicester, UK

The Youth, Young Adulthood and Society series approaches youth as a distinct area, bringing together social scientists from many disciplines to present cutting-edge research monographs and collections on young people in societies around the world today. The books present original, exciting research, with strongly theoretically- and empirically-grounded analysis, advancing the field of youth studies. Originally set up and edited by Andy Furlong, the series presents interdisciplinary and truly international, comparative research monographs.

Transitions to Adulthood through Recession
Youth and Inequality in a European Comparative Perspective
Edited by Sarah Irwin and Ann Nilsen

Youth, Technology, Governance, Experience
Adults Understanding Young Lives
Edited by Liam Grealy, Catherine Driscoll, Anna Hickey-Moody

Youth, Risk, Routine
A New Perspective on Risk-Taking in Young Lives
Tea Torbenfeldt Bengtsson and Signe Ravn

Rethinking Young People's Marginalisation
Beyond Neo-Liberal Futures?
Perri Campbell, Lyn Harrison, Chris Hickey and Peter Kelly

Youth in the Digital Age
Paradox, Promise, Predicament
Edited by Kate C. Tilleczek and Valerie M. Campbell

For more information about this series, please visit https://www.routledge.com/Youth-Young-Adulthood-and-Society/book-series/YYAS

Youth in the Digital Age

Paradox, Promise, Predicament

Edited by Kate C. Tilleczek and Valerie M. Campbell

LONDON AND NEW YORK

First published 2019
by Routledge
2 Park Square, Milton Park, Abingdon, Oxon OX14 4RN

and by Routledge
52 Vanderbilt Avenue, New York, NY 10017

First issued in paperback 2020

Routledge is an imprint of the Taylor & Francis Group, an informa business

British Library Cataloguing-in-Publication Data
A catalogue record for this book is available from the British Library

Library of Congress Cataloging-in-Publication Data
A catalog record has been requested for this book

ISBN 13: 978-0-367-58201-2 (pbk)
ISBN 13: 978-1-138-61312-6 (hbk)

Typeset in Times New Roman
by codeMantra

For the young people of today and tomorrow: teach us beyond technique!

Kate and Valerie dedicate this book to all of the young people from Australia, Canada, and Scotland who shared their time and spoke so freely with us about the digital age. Their thoughtful and insightful discussions about the digital age and its ecologies form the crux of this book and those to come. We have learned a great deal from these young people and their friends, and hope to continue our mutual conversations for years to come.

We also dedicate this book to Professor Andy Furlong, a co-investigator and a beacon of the study of young lives in global, social, political, and economic contexts. Andy is missed, yet his ideas rouse this book and those to follow.

Contents

List of figures

List of tables

Acknowledgements

As co-editors, Kate and Valerie acknowledge the many excellent researchers who worked with us to complete interviews and whose work on the *Digital Media and Young Lives* project made this book possible. UK interviewers: Cindy Corliss, Karen Young, Karen Cuthbert, Lisa Whittaker; Australia interviewers: John Smyth, Tim Harrison, Michael Hodgins; Canadian interviewers: Moira Ferguson, Natalie Baker, Ron Srigley, Matthew Munro, Janet Loebach, Monica Kelly. We are grateful to Rudy Madigan for her assistance with sourcing the technology to help us gather and work with the data and for the creation of the first draft of our Data Management Plan. Jonah Rimer (UK), Heather Barnick (Canada), and Ron Srigley (Canada) engaged with us in critical discussion and analysis of the data.

To all of the contributors (Heather Barnick, Brandi Bell, Janet Loebach, Matthew Munro, Jonah Rimer, and Ron Srigley), thank you for being prompt, engaged, and animated in the process of writing this book. Valerie appreciated your kind replies to her many emails. Kate is also grateful for the opportunity to work with this skilled team of researchers and scholars. Particularly, she is indebted to co-editor Valerie Campbell for her calm, cool, and collected approach to writing and editing. And to Ron Srigley for his critical insight about all things modern and his indelible writing talent.

Kate Tilleczek also thanks and acknowledges her brilliant colleagues and co-investigators in the project: Professor Katherine Boydell (Canada and Australia), Professor John Smyth (Australia), Dr Ron Srigley (Canada), and Professor Andy Furlong (Scotland). This project was enlightened by your ideas and unending care for scholarship and young people.

Valerie Campbell is grateful to Kate Tilleczek for her mentorship and for the opportunity to co-edit this book.

List of Contributors

Heather Barnick, Young Lives Research Laboratory, Canada, is a lecturer in the Department of Sociology and Anthropology at the University of Prince Edward Island. Her research interests include empathy as a tool for digital theorizing and methodology and MMORPG gaming and game design in China.

Brandi L. Bell is Assistant Director of the Young Lives Research Laboratory, Canada, and coordinates a variety of projects and initiatives focused on youth mental health. She holds a PhD in Communication Studies from Concordia University where she studied media representations of youth as social and political actors.

Janet Loebach, PhD, is an independent researcher and consultant whose work focuses on children's perception and use of their everyday environments, and the socio-environmental factors that influence youth behavior and well-being, including the role of digital media and technologies.

Matthew Munro is Senior Research Associate with the Young Lives Research Laboratory, Canada, and has a Master's of Applied Health Services Research from the University of Prince Edward Island. His research areas include youth access to mental health services, the history of youth anxiety, and the role of digital media in mental health.

Jonah R. Rimer, DPhil, School of Anthropology & Museum Ethnography, Oxford University, UK, is an anthropologist who specializes in digital and criminal phenomena. His Doctorate was completed in 2015 at Oxford University and focused on Internet sexual offending. He has since conducted research at the Young Lives Research Laboratory, Canada; Ryerson University; and Oxford University.

Ron Srigley, PhD, is a writer. His work has appeared in The Walrus, The Los Angeles Review of Books, and L'Obs, as well as in a variety of scholarly journals. He teaches philosophy and religious studies at Laurentian University and in the School of Liberal Arts and Sciences at Humber College, Toronto. He is the author of Albert Camus's Critique of Modernity, Eric Voegelin's Platonic Theology, and translator of Albert Camus's Christian Metaphysics and Neoplatonism.

Chapter 1

Young lives in the digital age

Kate C. Tilleczek

Introduction: a paradoxical age of youth

> A recent episode: students demonstrate and hold up the TGV in Angouléme [France] station. They pour down both sides of the train ... A shout here and a slogan there. An occasional outburst of rage – but against whom? It is as though they were barking at an artificial satellite. For the TGV is virtual reality ... an embodiment of speed, money and all those things that circulate – confronted here with *their* very real world of potential unemployment. A surrealistic confrontation between time's arrow and a youth already spent. (Baudrillard, 2002, p. 57)

This account from "The Powerlessness of the Virtual" in Jean Baudrillard's *Screened Out* encourages us to think about new forms of inequality arising from technological developments with which youth must contend. A youth demonstration in France against the operation of the Train à Grande Vitesse (TGV), the fastest train on earth at 322 km/hour, illustrates the complex ways in which young people approach and live with modern technology. They are not only using digital tools (e.g., cell phones, texting, social media, and gaming) but also considering and reacting to macro-facets of the technological world that affect so dramatically the way they live, work, learn, love, and enjoy themselves. Baudrillard's train episode points to virtual techno-reality, against which French youth protest in vain, while their power and affluence are mainly excluded. They trail behind this emblematic sign of technological progress, alternately protesting against it and scrambling to purchase the latest digital inventions of which the TGV is a powerful symbol.

There are weighty consequences for youth, no matter how they act. For instance, as labour becomes automated and robots do ever-greater shares of human work, youth unemployment and underemployment has increased, making more young people marginal to global economies (Carr, 2014; Kelly, 2017; MacDonald, 2017). Nearly 75 million youth are unemployed around the world, an increase of 4 million since 2007, and there is no indication of a positive change in sight (International Labour Office, 2012). The global youth unemployment rate hovers around 13% and dramatically overshadows the adult unemployment rate of 4% (UN Department of Economic and Social Affairs, 2018). Youth unemployment has increased social inequality by 4% points in Western countries and by 8% points in specific countries, such as Greece, Ireland, Italy, Portugal, and Spain (Matsumoto, Hengge, & Islam, 2012). The position of youth in society is confounded by growing income inequality as they continue to be more disadvantaged relative to adults (OECD, 2011; UNICEF, 2016; Wilkinson & Pickett, 2009).

The digital age is meant to save young people from these burdens, but late modernity is characterized by weakening of the social networks that traditionally supported young people (Furlong, 2012). Pathways and trajectories to flourishing adulthoods are now multidimensional, non-linear, and complicated. The social, economic, and political problems youth encounter most often are now seen as individual failings, solved only through personal action (Furlong, 2017; Furlong & Cartmel, 2007). A key feature of the digital age is an entrenched fallacy of individualism and individual responsibility and control which acts to (a) obscure social relations between people and institutions, and (b) suggest that we solve our problems individually. This epistemological fallacy of late modernity (Furlong, 2012) is an important touchstone for analysis in this book.

This paradoxical age of youth includes the fact that there are now 1.8 billion young people (10–24 years) on the planet, the largest youth cohort in history, with 90% living in low- and middle-income countries (Patton et al., 2016). Digital technologies are seen by many to play a potentially fundamental role in youth well-being and linked to the success of meeting the United Nations' Sustainable Development Goals (Patton et al., 2016). However, while the promises of digital technologies appear to be vast, they are also being questioned in North America, where neuro-scientific advances show adolescence as a period of dynamic brain development in which interactions with social environments fundamentally shape experience and capabilities (Patton et al., 2016). As we report in detail in Chapter 3, there is mounting evidence of threats to youth well-being through declining mental health, cognitive and moral capacities, and loss of human contact and relationships (Pew Research Center, 2015; Tilleczek & Srigley, 2017; Twenge, 2017), and "of exacerbating forms of exclusion, amplifying pre-existing risks, or advancing state surveillance and/or commercial exploitation" that disempower youth (Livingstone, Nandi, Banaji, & Stoilova, 2017, p. 1).

The book attends to the paradox and to scholars; commentators; governments; and, increasingly, technology industry ex-patriots who acknowledge these concerns. It also poses questions as to whether (or how) governments and the digital technology industry are able to support the well-being of young people when there is so much money to be made from youth who are considered the future of progress for techno-science. Therefore, the book engages a most complex narrative about the forms and consequences of digital technology, and how young people enact the interplays between themselves, digital tools, and their ways of living. This is the perplexing digital age in which the young people with whom we spoke are living. As Easton (17)[1] stated, "I think humans are going to become the new technology and companies are going to be selling upgrades to people." And Naomi's (20) perspective is reported in Chapter 3 as a reminder that

> most of their apps and social media apps are geared towards our age group because I feel like you [sic] can do the most ... I don't know why, it feels like they want to make us do damage. I don't know who they even is [sic], but I feel like we're just the most vulnerable crowd for them to zone in on and for them to get as much as they possibly can out of us for their benefit.

Castells's (2015) positioning of the oscillation between outrage and hope in the networked society is useful to our analysis. Throughout the book we point to the potential for power and agency to arise from being networked and to the affordance of unique types of resistance in the digital age (a theme taken up in detail in my forthcoming book). But these young people also feel acute despair about control, surveillance, and loss of privacy as the digital age intensifies individualism through breakdowns in social cohesion, human connection, and the rapacious collection and sale of their data and selves for profit. As Aaron (18) suggested for forging a better future, "Don't spend most of your life on the Internet, there's much more to live for than just sitting on the Internet." There is an emerging state of broken trust between these young people and the Big Tech[2] companies that control the tools and forms of life in the digital age. The current tech-lash is fundamentally about this loss of trust and potential for young people to capitalize and flourish, issues we detail in the book and in context of the shifting tides of Big Tech's doings. Their concerns about both are illustrated in this book and point to the ways in which well-being is becoming compromised.

We scaffold the concerns of Furlong (2011) that social, political, and economic conditions and analyses are becoming obscured in late modernity such that the digital age brings forth "the same sort of conditions that are linked to processes of social reproduction in capitalist societies everywhere" (Furlong, 2011, p. 5). We therefore speak to these young people about the inequalities that exist in their digital ecologies (tools and ways of life) and place social structural analyses alongside interdisciplinary literatures that contend with them. Slate's Glaser helps to summarize the newest tech-lash perspectives:

> if you think Facebook is worth deleting over its issues, then call your elected official to regulate the company, as well as other companies, like Google, that profit from harvesting our personal details to sell ads tailored to us across the Internet. Because it's becoming clearer that we can't trust these companies to regulate themselves. (Glaser, 2018, para. 12)

This is the flavour of conundrum in the digital age, with its tech-loves and tech-lashes. It is difficult to comprehend how quickly the pendulum has swung in the first 18 years into this new millennium. The young participants in our study were, on average, 16–17 years of age in 2013–2014, and were chosen particularly as part of the group that lived through this surge in time spent online and with increasing forms of digital media and connectivity. They saw the shift to mobile technologies and rapid increases in design of social media platforms, apps, and tech gear (mobile phones, Google glasses, and so forth). Since they were children in the year 2000, these young people moved online and behind screens in growing numbers and in new ways. Internet users increased worldwide from less than 500 million in 2000 to over 3 billion in 2015 (Murphy & Roser, 2018). By far, the majority of Internet users are young people; "the proportion of young people aged 15–24 using the Internet (71%) is significantly higher than the proportion of the total population using the Internet (48%)" (ITU Telecommunication Development Bureau, 2017). By 2007, when the first iPhone launched (Ritchie, 2018), these young people were coming of age to witness the burgeoning of the total mobile 24/7 ubiquity of digital media. However, until 2015 it was difficult to raise a serious discussion or debate about the meaning of digital media for the lives of youth. Outside the social sciences, there were touchstone and long-standing critiques of the humanities (see Srigley, 2011; Tilleczek & Srigley, 2011, 2017 for reviews). But journalists, educationalists, and technology industries were still engaged in their love-in with digital technology. In universities and across school boards, administrators and educators were lining up with Big Tech and their spin doctors to procure and dole out the latest technological solutions to the problems that ail education, literacy, and youth engagement. SMART Boards; clickers; beepers; Technology, Entertainment, Design (TED) talks; iPads; and Twitter feeds filled the sacred space of the classroom. Any mention of Marshall McLuhan or Jean Baudrillard's (or others') critiques of technology was met with dismissive assurance: "yes, but technology is just a neutral tool." Some of the youth participants also saw technology as "just tools," such as Gabriel (20), "technology is a tool that makes things better or easier to do," and Harper (17), "it is a tool easy for us to easily … access so much information. It makes our lives easier." However, the neutrality of the tool was a burgeoning question for some, such as Violet (17): "But my anger is more like towards the people who leak the nudes and not so much technology. Because it's just a tool, you know, it's not the source of all evil." For other young people still, the neutrality of technology and its designs was always in question. As Coke machines and techno-gadgets filled public schools and universities, the story of Big Tech and the technological worldview of which it was an essential part remained largely untold. This book tells some of that story from the point of view of young people who lived through this time.

Indeed, we know that the treatment and lives of young people continually alters with social, political, economic, and technological shifts (Furlong, 2017; James & Prout, 2015). It is worth pondering the fact that in wartime youth were seen as more competent and ready for the front lines of battle, while during economic depression they were infantilized since they stand in the way of an already cluttered labour market (Enright, Levy, Harris, & Lapsley, 1987). Today youth are alternately seen as activists, consumers, citizens, and digital natives. They are thought to be comfortable and savvy about digital media at the same time that they are becoming tangential to and precarious in the labour market, and in need of increasing years of education to compete in and for the good of a technocratic society (Kelly, 2017; MacDonald, 2017). Those who see great benefit in digital technological advances see youth as knowledgeable, willing, and able participants, and warn us against the demonization of digital media (boyd, 2015; Odgers, 2018). Those who offer wary critiques of technology and digital media see youth as contained in a labyrinth of gadgetry and consumerism made addictive by an ever-expanding and powerful technology industry, part of the new cyber-proletariat (Dyer-Witheford, 2015). In this expanding field of study, youth are both valorized and demonized, in turn. The dialectic of representation and treatment of youth is inadequate for apprehending the digital age and ways in which youth actively and potentially negotiate it for their well-being and write their futures.

The design of the book

This book is based on a project that unpacks profound conundrums and dialectics of young lives in the digital age. We report on the methods and emerging themes of my Social Sciences and Humanities Research Council of Canada (SSHRC)-funded *Digital Media and Young Lives over Time and Place* project (hereafter referred to as the *Digital Media and Young Lives* project). It is worth noting that SSHRC is a Canadian governmental funding body with no ties to the technology industry, based upon peer review processes, and mandated to foster open debate and scholarship. I welcomed the invitation to design and execute a research project with an international and interdisciplinary team, using a humanities-infused imagination and methodological frankness. In Chapter 2 we detail the methodological and ethical decisions we made as we tried to "find new ways of integrating empirically grounded and dialogical strategies of youth research within interdisciplinary and theoretically sophisticated frameworks of comparative analysis" (Cohen & Ainley, 2000, p. 91). We describe a methodological openness with which the team indulged lessons from Paul Feyerabend (1993) and C. Wright Mills (1959). This book presents data from a youth-attuned qualitative research design (Tilleczek, 2018), which included video-recorded face-to-face interviews with 105 youth participants, who also completed a short demographic survey and shared their digital media activity on platforms such as Facebook, Instagram, Twitter, Pinterest, and Tumblr. Forty of these young people also returned for an additional on-film interview with their "digital shadows," who were friends with whom they share online experiences. We heard from these 40 young people on two occasions. Their digital shadows were freely chosen and invited by the young participants and provide lively two-participant conversations that added depth to the data, leading to 40 additional interviews (dyadic). Thus, we recorded the voices of 185 young people, from which we draw conclusions and analysis in tandem with interdisciplinary literatures.

The project and book are team efforts including collaborative analysis and writing of these chapters and further forthcoming publications. This book is designed to present youth perspectives on technology (predicaments), the potential benefits they might derive from technology (promises), and the mixed and conflicted usage and the meanings of technology within a rapidly shifting digital age (paradox). There were remarkable narratives and messages that jumped out across conversations and scholarly analysis. Our objective is to provide insight about how young people actively cope with the practicalities of digital technology in their daily lives. What are the pressing challenges and opportunities? How are youth actively responding to, resisting, and negotiating that which they find problematic? Chapter 3 concretizes challenges and possibilities in the context of the landscape of literatures about youth well-being (social, emotional, physical, and spiritual). It begins with a review of competing models of youth well-being and advances our own holistic model based upon the hopes and challenges that these young people expressed. We pay specific attention to concerns raised and strategies employed for navigating them. Notably, we uncover the need for a new domain of well-being (we call this domain "digital lives") and suggest it to those who are trying to understand youth today. This domain includes themes of managing the tensions/pressures of online life, balancing daily use to better effect, and navigating shifting digital ecologies. We posit additional potential ways in which measures and models of well-being can be enhanced across seven domains: health, relatedness, equity, education/work, engagement, affordable living, and space/environment of well-being (UNICEF Canada & Students Commission of Canada, 2017). Our graphically illustrated holistic model (Figure 3.1) builds on young people's concerns about distraction, lack of attention, violence, becoming less physically active, and losing the joys of nature. The timely, simultaneous development of the United Nations International Children's Emergency Fund (UNICEF) Canada's new *Child/Youth Well-Being Index* ([CY-Index], UNICEF Canada, n.d.) is a backdrop to interpretation of our analysis in Chapter 3.

Chapters 4 through 7 then detail specific aspects of the paradoxes, promises, and predicaments. For instance, Chapter 4 investigates privacy and surveillance as central ethical concerns for these young people. We discuss the actions of Edward Snowden and more recent debates about the power of social media, such as the Cambridge Analytica scandal and Big Tech's complicity in surveillance, control, and privacy. This context opens onto comparisons about the ways in which privacy and surveillance concerns were discussed by young people. They addressed them in micro- and macro-political ways that illustrate processes enabling state and corporate interests to colonize cyberspace. Our young participants from Canada, Scotland, and Australia spoke about how they attempted to guard their privacy while knowing that total resistance was futile. The dialectical pulls of concerns about privacy and control alongside requirements to maintain an online presence illustrate how technology and capitalism meet in insidious ways. To wit, modern society is leaving young people to navigate deep dilemmas which confront all of us – namely to weigh personal privacy and civil liberties against promises of a technological worldview and its tools. Chapter 5 heightens insight into this tall order with an intriguing "no phone experiment" and is co-written by a philosopher and professor who studies critiques of modernity. A philosophical lens was added to analysis when his students unplugged from their cell phones and wrote heartfelt essays about the experience. These university students were different from those we interviewed. But their essays reverberate with similar concerns about human relationships, freedom, productivity, morality, engagement, parents, and safety. The university students in the "no phone experiment" also felt trapped in the digital age.

Such traps are those that young people attempt to overcome with their "digital capital." Chapter 6 explores this emerging concept as one conceivable manner for potential technological tool use that might avoid the utopian worldview that goes with the design, sale, and use of digital media. Digital capital does not easily emerge for youth, given the larger problem of their precariousness in the digital age. But there are daily ministrations and actions taken by youth to enact an *interplay* between digital tools and the social and economic relations that surround them. In this interplay, young people demonstrate how symbolic, social, cultural, and monetary digital capital might be gained. They pose fundamental questions as to what end, and for how long, their digital capital has been gained.

Finally, Chapter 7 deeply examines how social relationships fit into the complex mix. How are new forms of mediated sociability experienced and navigated? Led by an anthropologist, Chapter 7 sets a context in media and commentator concerns about distraction and loss of human connection in the digital age, with its new social and interactional norms, practices, and connections when human communication is mediated by technology. It should be no surprise that deep tensions exist at the level of human relationships. Online sociality, social norms, communication, and interaction are discussed by youth. The motifs they share are dialectical: *easy reach vs difficult distance, forging connections vs changing relationships,* and *enhanced vs suffering sociality.* Young people experienced and perceived multiple positives, they were able to better organize with others, they had enhanced communication and the ability to transcend place, and many found belonging and support in their online connections. Real concerns about anonymity, exclusion, misrepresentation, narcissism, and anxiety are peppered throughout these online human relationships. There is no doubt that sociality in the digital age is complicated and contradictory, and these young people are up to the tasks of attempting to negotiate emerging social spaces.

The conclusion of the book summarizes these profound conundrums as they appear across the many aspects of young lives and digital ecologies presented here. Chapter 8 returns us to a place in which this introduction began: namely with images that are just as emblematic as Baudrillard's train – important scenes to which we have become privy through this project. Particularly, the train illustrates the quickly moving target that is the ecology of the digital age (its tools and ways of living). To what extent are young people keeping up? What does it mean for their well-being? Do youth protest in vain? Where is their power and affluence in developing digital capital? Who is writing their future in the digital age, and to what end? What will be the place of the world's largest-ever cohort of young people in writing this future? These gracious young people from Canada, Australia, and Scotland have provided important momentum for answering such questions. It is my hope that readers will embrace that which has been shared by these youth and come to understand the joys and deep tensions with which they are living. Together we could forge a kinder and gentler place *with, for,* and *by* youth. Their early clues for a good way forward are calls to be seriously responded to.

> It's kind of, there's always a risk and a disadvantage with everything. Technology would be a great enhancement for humankind, but also it would be probably the downfall of it. Because if everything's suited to technology, someone can hack in and, bam! Either your lives are ruined or the whole society is ruined. So it's kind of scary, but it's also hopeful. (Jaxon, 20)

To hear directly from some of these young participants, view the digital portraits (see Chapter 2 for a description) featuring:

Cameron (https://www.youtube.com/watch?v=hoBBB9sWfHU)
Cooper (https://www.youtube.com/watch?v=zcSm4RkD0Fk)
Willow (https://www.youtube.com/watch?v=78nQwDew5U0)
Sebastian (https://www.youtube.com/watch?v=PbO7ld8Kc64)

Notes

1 All youth names are pseudonymous. We provide only the age at Phase I of the study, not country, gender, ethnicity, or employment status.
2 Facebook, Google, Apple, Amazon, and Microsoft (Lotz, 2018).

References

Baudrillard, J. (2002). *Screened out* (C. Turner, Trans.). London, UK: Verso.

boyd, d. (2015). Social media: A phenomenon to be analyzed. *Social Media Society, 1*(1), doi:10.1177/2056305115580148

Carr, N. (2014). *The glass cage: How our computers are changing us.* New York, NY: WW Norton.

Castells, M. (2015). *Networks of outrage and hope: Social movements in the internet age* (2nd ed.). Cambridge, UK: Polity Press.

Cohen, P., & Ainley, P. (2000). In the country of the blind? Youth studies and cultural studies in Britain. *Journal of Youth Studies, 3*(1), 79–95. doi:10.1080/136762600113059

Dyer-Witheford, N. (2015). *Cyber-proletariat: Global labour in the digital vortex.* London, UK: Pluto.

Enright, R. D., Levy, V. M., Harris, D., & Lapsley, D. K. (1987). Do economic conditions influence how theorists view adolescents? *Journal of Youth and Adolescence, 16*(6), 541–559.

Feyerabend, P. (1993). *Against method* (3rd ed.). New York, NY: Verso.

Furlong, A. (2011, Dec). *Digital capital and inequality in late modernity.* Paper presented at The Future of Youth Sociology Symposium at the Australian Sociological Association Conference, Sydney, Australia.

Furlong, A. (2012). *Youth studies: An introduction.* London, UK: Routledge.

Furlong, A. (2017). The changing landscape of youth and young adulthood. In A. Furlong (Ed.), *Routledge handbook of youth and young adulthood* (2nd ed., pp. 19–27). Abingdon, UK: Routledge.

Furlong, A., & Cartmel, F. (2007). *Young people and social change: New perspectives.* Buckingham, UK: Open University Press.

Glaser, A. (2018, Mar 21). The problem with #DeleteFacebook. *Slate.* Retrieved from https://slate.com/technology/2018/03/dont-deletefacebook-thats-not-good-enough.html

International Labour Office. (2012). *Global employment: Trends for youth 2012.* Geneva, Switzerland: International Labour Organization. Retrieved from http://www.ilo.org/wcmsp5/groups/public/—dgreports/—dcomm/documents/publication/wcms_180976.pdf

ITU Telecommunication Development Bureau. (2017). *ICT facts and figures 2017.* Retrieved from https://www.itu.int/en/ITU-D/Statistics/Documents/facts/ICTFactsFigures2017.pdf

James, A., & Prout, A. (2015). *Constructing and reconstructing childhood* (3rd ed.). London, UK: Routledge.

Kelly, P. (2017). Young people and the coming of the third industrial revolution. In A. Furlong (Ed.), *Routledge handbook of youth and young adulthood* (2nd ed., pp. 391–399). Abingdon, UK: Routledge.

Livingstone, S., Nandi, A., Banaji, S., & Stoilova, M. (2017). *Young adolescents and digital media: Uses, risks and opportunities in low-and middle-income countries: A rapid evidence review*. London, UK: Gage.

Lotz, A. (2018, Mar 23). 'Big Tech' isn't one big monopoly – it's 5 companies all in different businesses. *The Conversation*. Retrieved from https://theconversation. com/big-tech-isnt-one-big-monopoly-its-5-companies-all-in-different-businesses-92791

MacDonald, R. (2017). Precarious work: The growing précarité of youth. In A. Furlong (Ed.), *Routledge handbook of youth and young adulthood* (2nd ed., pp. 156–163). Abingdon, UK: Routledge.

Matsumoto, M., Hengge, M., & Islam, I. (2012). *Tackling the youth employment crisis: A macroeconomic perspective*. (Employment Working Paper No. 124). Geneva, Switzerland: International Labour Organization. Retrieved from http://www. ilo.org/wcmsp5/groups/public/—ed_emp/—emp_policy/documents/publication/ wcms_190864.pdf

Mills, C. W. (1959). *The sociological imagination*. New York, NY: Oxford University Press.

Murphy, J., & Roser, M. (2018). *Internet*. Retrieved from https://ourworldindata. org/internet

Odgers, C. (2018). Smartphones are bad for some teens, not all. *Nature, 554*, 432–434. Retrieved from https://www.nature.com/articles/d41586-018-02109-8

Organisation for Economic Cooperation and Development, [OECD]. (2011). *Against the odds: Disadvantaged students who succeed in school*. OECD Publishing. doi:10.1787/9789264090873-en

Patton, G. C., Sawyer, S. M., Santelli, J. S., Ross, D. A., Afifi, R., Allen, N. B., ... Bonell, C. (2016). Our future: A lancet commission on adolescent health and well-being. *The Lancet, 387*(10036), 2423–2478.

Pew Research Center. (2015). *Internet seen as positive influence on education but negative on morality in emerging and developing nations*. Retrieved from http:// www.pewglobal.org/files/2015/03/Pew-Research-Center-Technology-Report-FINAL-March-19-20151.pdf

Ritchie, R. (2018, Jan 9). 11 years ago today, Steve Jobs introduced the iPhone. *iMore*. Retrieved from https://www.imore.com/history-iphone-original

Srigley, R. D. (2011). *Albert Camus' critique of modernity*. Columbia, MO: The Curators of the University of Missouri.

Tilleczek, K.C. (forthcoming). *Young cyborgs? Youth resistance to technology*. Toronto, ON: University of Toronto Press.

Tilleczek, K. (2018). Qualitative methodology in adolescent research. In S. Hupp, & J. Jewell (Eds.), *The encyclopaedia of child and adolescent development*. New York, NY: Wiley.

Tilleczek, K., & Srigley, R. (2011). Modern youth at work and play. In K. Tilleczek (Ed.), *Approaching youth studies: Being, becoming and belonging* (pp. 66–86). Toronto, ON: Oxford University Press.

Tilleczek, K., & Srigley, R. (2017). Young cyborgs? Youth in the digital age. In A. Furlong (Ed.), *Routledge handbook of youth and young adulthood* (2nd ed., pp. 273–284). New York, NY: Routledge.

Twenge, J. M. (2017). *IGen: Why today's super-connected kids are growing up less rebellious, more tolerant, less happy–and completely unprepared for adulthood–and what that means for the rest of us*. New York, NY: Simon and Schuster.

UN Department of Economic and Social Affairs. (2018). *The sustainable development goals report 2018.* Retrieved from https://unstats.un.org/sdgs/report/2018

UNICEF. (2016). *The state of the world's children 2016: A fair chance for every child.* New York, NY: United Nations Children's Fund. Retrieved from https://www.unicef.org/publications/files/UNICEF_SOWC_2016.pdf

UNICEF Canada. (n.d.). *Measuring well-being.* Retrieved from https://www.unicef.ca/one-youth/child-and-youth-well-being-index/

UNICEF Canada & Students Commission of Canada. (2017). *My cat makes me happy: What children and youth say about measuring their well-being.* Toronto, ON: UNICEF Canada. Retrieved from http://www.unicef.ca/sites/default/files/2017-08/UNICEF_One%20Youth%20Report.pdf

Wilkinson, R., & Pickett, K. (2009). *The spirit level: Why equality is better for everyone.* New York, NY: Penguin.

Chapter 2

Methods and ethics *with, for,* and *by* youth in the digital age[1]

Valerie M. Campbell, Kate C. Tilleczek, and Janet Loebach

Introduction

We have argued elsewhere (Tilleczek & Srigley, 2017) that the task of understanding the lives of young people in the digital age is a complex and high-stakes business. To even attempt this understanding requires a humanities-infused imagination and a liberal methodological freedom; "we have to find new ways of integrating empirically grounded and dialogical strategies of youth research within interdisciplinary and theoretically sophisticated frameworks of comparative analysis" (Cohen & Ainley, 2000, p. 91). This is not to say that adherence to method is without merit. Methods are indeed valuable features of inquiry but should be subordinate to the phenomenon of study and open to fluctuation and imagination. It is this methodological openness that we bring to our study of the lives of young people in the digital age with its rapidly shifting sets of tools, media, and practices. The goal of our approach is to follow scholars of inquiry such as Feyerabend (1993) and Mills (1959) and engage experiences, ideas, and practices of young people as they live, work, learn, and play in digital places. The approach to inquiry we have developed attempts to facilitate the achievement of that goal. Described in Chapter 1 of this book and in Tilleczek and Srigley (2017), our project and methodological framework was born of an interdisciplinary conversation that traces historical, economic, social, and philosophical aspects of modern technology's aims. The Social Sciences and Humanities Research Council of Canada (SSHRC)-funded project was proposed and generated in Tilleczek's *Young Lives Research Laboratory* (YLRL)[2] in Canada. It began in understanding diverse traditions that have offered critiques of the modern technological project as a way to avoid the streamlined, reductionist, and paltry explications that we had grown accustomed to.

The project and its methodology were guided by one main question: How are the everyday lives of young people influenced by digital media and over time and place? Supplementary questions included: What are the promises and problems encountered by young people when they engage with digital media? How is digital media involved in youth agency and resistance? What are the social and educational impacts? Is an emerging form of digital capital arising? If so, how are young people using it? We examined these areas of inquiry by engaging with young people through a variety of interview conversations: What are the benefits and/or constraints of technology for you? How does it operate in your life (including social relationships and school)? What are the economic consequences of participating or not participating in the digital age? Where do you go online and what does that mean for you? How do you present yourself online?

This chapter provides detail into the modes of inquiry and decision-making processes that were designed to address these questions. The proposal and project pointed out gaps in the study of youth and digital media such as (a) longitudinal research, (b) research about and with marginalized youth, (c) comparisons within and across diverse youth cultures, and (d) international comparisons of digital media use and impacts. To these ends, the *Digital Media and Young Lives* project was designed as a longitudinal, cross-cultural study spanning five years. The research team was international, with investigators in Canada, Scotland, and Australia. We planned two phases of data collection, involving a diverse range of youth from those countries, supporting both inter- and intra-cultural comparisons. This book addresses early emerging themes from Phase 1 data: well-being, privacy, resistance, digital capital, social relationships, and education. Areas of focus for future dissemination of the project data include the experiences of Indigenous participants, longitudinal and cross-cultural comparisons, and analysis of the social media data. Phase 1 findings have also been reported in other venues (e.g., Srigley & Tilleczek, 2016; Tilleczek, 2017a, 2017b, 2018a, forthcoming; Tilleczek & Srigley, 2017; Tilleczek & Tilleczek, 2016).

This chapter guides the reader through our sets of decisions and practices of inquiry: (a) To whom should/could we speak and how? (b) What are the ethical dilemmas considered/encountered? (c) What do the data mean? (d) How can we most effectively analyze and share what we have learned? We conclude with a summary of our modes of inquiry, lessons learned for modifications, and new directions for the study of the digital age *with*, *for*, and *by* young people. Throughout the chapter, we explicate our decision-making processes and methodological adaptations. We began with a plan of action and a set of research tools yet as the project progressed, we continuously evaluated our methods, discussing and making changes as appropriate while remaining focussed on the research questions and committed to ethical practice.

To whom should/could we speak and how?

As we began developing research tools and protocols we considered potential data sources, *which* young people would be best positioned to speak to the research questions? Where would we find them and how would we connect with them?

With whom should we speak?

We wanted to speak with youth who had the opportunity to be involved in digital media but could also reflect on a time when it was less pervasive. Thus, we began our recruitment with youth aged 16–19 years, selected because youth within this group are likely to be fully immersed in the contemporary digital world (boyd, 2007) yet old enough to recall and reflect on life before digital technology was ubiquitous (James et al., 2010; Livingstone & Brake, 2010). Indeed, youth in this age group would have a unique perspective as they were entering adolescence just as the social media explosion was beginning. The longitudinal nature of this project expanded our age range to 24 years (over the five years of the study) and provided an opportunity to explore how and why digital media use had, or had not, changed over that span.

The original sampling frame of youth participants (Table 2.1) was purposive and was based on the principles of maximum variation (Tilleczek, 2008) to recruit according to categories of theoretical interest. The frame was initially designed to recruit participants within each of the three study countries, and across three diverse subcategories, to maximize the range of youth perspectives collected. The initial subcategories targeted for recruitment in each of the study countries were third generation, newcomer, and Indigenous youth.

For the purposes of this study, "third generation" youth were defined as those born in the study country and representative of those well embedded in its social and cultural life. "Newcomer" youth were defined as youth not born within the study country, but ideally having arrived there within the previous five years so they may still recall experiences from their previous locale or culture. Newcomer youth may reflect different experiences than those youth who have lived in the country their whole lives and are of interest for several reasons: (a) there is little published about newcomers' relationship to digital media; (b) these young people are often socially marginalized by virtue of being a visible minority, having language difficulties and/or poor social status; and (c) their stories contain elements of global digital media experiences for cross-national comparison. Youth were considered to be Indigenous if they self-identified as such.

There is a continued debate among qualitative researchers around the concept of "enough" participants, described in a review by Baker and Edwards (2012). While some suggest rules of thumb for repetitions (McCracken, 1988; Patton, 1990), those concepts remain vague and less than helpful. Hennink, Kaiser, and Marconi (2017) explored the concept of saturation in depth with mixed results; the primary determinant for "enough" was the goals of the study rather than a fixed number. The 14 scholars surveyed by Baker and Edwards (2012), including "seasoned methodologists and early career researchers" (p. 3), provided responses that ranged from 1 to 60; some spoke of data saturation as a guideline, while there were others who simply responded: "it depends" (p. 3). Dworkin (2012) suggested that "it depends" was a reasonable answer and the rigour of the methods should be considered, although she did recommend 25–30 as sufficient to achieve saturation. We considered the debate outlined in the literature as well as our resources and the diversity of our participants; we engaged in our own open debate on sample size, and ultimately agreed that a minimum of eight participants were to be recruited in each subcategory. Further team discussions revealed that there was no distinct Indigenous population in Scotland, and the Australian team did not have sufficient connections with local Indigenous groups to support necessary connections with the community. It was therefore determined that the Indigenous category would be oversampled in Canada with no comparative participants from the other two study countries. Due to the greater resources and connections of the Canadian team, it was also determined that twice as many youth participants would be selected in each of its subcategories. The original participant goals therefore were to engage a minimum of 96 young people at both Phase 1 and Phase 2, across six subcategories (see Table 2.1).

Youth participants were also asked to select a *digital shadow*, conceived for this project as a close friend with whom the youth in our study conversed daily, whether online or offline, to participate with them in a second, dyadic, interview. Digital shadows were important contributors to the study, allowing for deeper exploration of the effect of digital media on youth relationships and communication. Previous research suggested that young people may feel more comfortable and empowered by the inclusion of their friends in the research process (Bell & Campbell, 2014). A dyadic interview can prompt a more open and engaging conversation in which the participant and their digital shadow together discuss the role digital media plays in each other's lives, and the similarities and differences in their use of and perspectives on digital technologies. Dyadic interviews also allowed the researchers to observe the role digital media plays within a friendship, helping to answer questions regarding social interaction

Table 2.1 Original Project Participant Goals

Category	Canada	Scotland	Australia	Total
Third generation	16	8	8	32
Newcomer	16	8	8	32
Indigenous	32	0	0	32
	64	16	16	96

and relationships in a technological world (Bell & Campbell, 2014). Digital shadow narratives scaffold the trustworthiness of both field work and analysis by providing additional youth perspectives of digital media that can be checked against those of their friend, and by adding depth to the discourse produced during the initial individual interview (Morgan, Ataie, Carder, & Hoffman, 2013).

The inclusion of a digital shadow for each original participant in effect doubled the sampling frame, increasing the target participant total from 96 to 192 youth. While the original sampling frame would provide some diversity among participants, during the initial face-to-face meeting of the full team we recognized the need to reflect an even broader range of young lives, including diversity in terms of gender, education, and employment status within each youth category. An amended sampling frame was therefore developed to include three additional categories: In Education (currently attending an educational institution or enrolled and planning to return in the fall); In Employment (currently employed or participating in a work training or apprenticing programme); and those in neither education, employment, nor training (NEET) (Furlong, 2006). With our goal of a minimum of eight participants per subcategory, this significantly increased our participant numbers. Early into the youth interviewing phase, however, it was apparent that many youth were both working and in an educational programme; these two subcategories were therefore collapsed into a single category with a minimum of eight participants (before the inclusion of digital shadows). The sample goal was therefore amended to a total of 80 youth participants for Canada and 32 participants each for Scotland and Australia for a total of 144 youth, an additional 48 participants over the original sampling frame. The inclusion of digital shadows offered the potential to double the participant total to 288 participants, 160 youth for Canada and 64 each for Scotland and Australia.

Gender was not added into the amended frame, but recruitment efforts attempted to achieve gender balance. The question of gender was open-ended: "How do you identify your gender?"; our participants identified as either female (58%) or male (42%). As we were recruiting, the team also attempted to note where young people fell within the digital continuum, from daily users to youth who rarely, if ever, utilize digital media or technologies, to attain a balance of perspectives. One hundred nineteen participants completed a survey (Facesheet, described later) on their daily social media activity. Figure 2.1 shows the continuum of use based on reported hours/day. The chart is based on the five most used social media as reported on the Facesheets: Facebook, Twitter, Instagram, texting, and email. Usage ranges from one hour/day to 23.5 hours/day. Participants often had multiple apps open at one time, which accounts for the inflated top number.

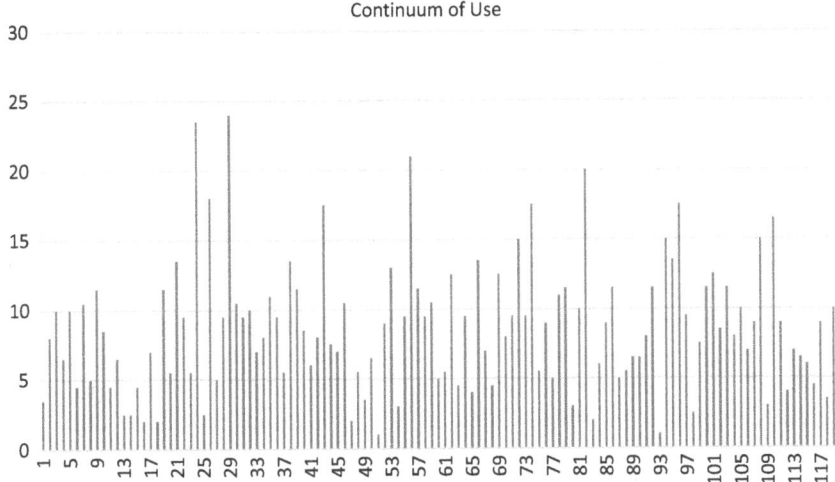

Figure 2.1 Continuum of Social Media Usage.

Locating youth with whom to speak

We drew upon our existing contacts in Canada, Scotland, and Australia to select sites for recruitment. The Canadian participant sample was drawn from three distinct areas of the country: eastern Maritime Provinces (Prince Edward Island (PEI) and Nova Scotia), urban Ontario (Toronto), and rural communities in Northern Ontario. Scotland participants were recruited in Glasgow. In Australia, youth participants came from two research sites: Ballarat and Sydney. The research team at each study site developed their own context-specific version of the study recruitment tools and methods but all used the same interview protocol.

Ethics approval was sought and received from the University of Prince Edward Island (UPEI) Research Ethics Board for the Canadian, Glasgow, and Sydney portions of the study. The participants in these locations were recruited and interviewed by researchers employed directly through UPEI so no other institutional approval was required. The University of Ballarat provided ethics approval for the Ballarat data collection. The Mi'kmaq Confederacy of Prince Edward Island (MCPEI) approved our request to recruit Indigenous youth from PEI. The team worked with MCPEI and the Native Council of PEI (NCPEI) in their recruitment efforts; these organizations helped to put us in contact with local Indigenous youth or youth leaders. While we were not able to directly recruit Indigenous youth in Nova Scotia or Northern Ontario, some participants from these regions did self-identify as Indigenous. Indigenous participants comprised 17% (*n*=25) of the total number of participants.

As recruitment progressed in each study country, the team made several changes to better connect with the youth populations; these revision strategies were first vetted and approved by the UPEI Research Ethics Board. The changes allowed the team to use digital networks, such as Facebook, Twitter, and Reddit, to recruit and then communicate with potential participants, and allowed for the possibility of conducting interviews via Voice over Internet Protocols (VoIPs) such as Skype or Adobe Connect. After piloting the recruitment of digital shadows for participation it was also determined that the most effective strategy was to turn this over to the primary participant who would contact and arrange for their friend to participate in a dyadic interview.

In Canada, recruitment posters and information handouts were provided to each organizational contact. In a few cases, members of the Canadian research team were invited to make presentations to youth groups to recruit participants. Researchers in the Maritime provinces utilized a shared recruitment document that allowed all team members to add and view new youth-related contacts and the status of their recruitment work, which helped to avoid duplicate efforts. Researchers in Toronto primarily worked with Concrete Roses, a Youth Services organization in downtown Toronto (http://concreteroses.ca/) to recruit participants. In Northern Ontario, recruitment techniques included the use of Facebook, snowballing techniques, and direct contact with youth at the public library and youth drop-in centres.

Members of the research team in Glasgow and Sydney received recruitment materials from the Canadian team which they adapted for their specific context. Their primary recruitment methods were the placement of posters in areas youth were known to congregate (coffee shops, youth centres, etc.), presentations to local youth through various service organizations, social media, and snowballing techniques. The Ballarat team recruited through existing networks of youth and youth centres (Smyth & Harrison, 2015).

To encourage participation, all participants were given a *Certificate of Volunteer Participation* issued by the YLRL (Canada). The Certificate outlined the research project and detailed the importance of their contributions to the study; it was prepared in advance and signed by the Principal Investigator (Dr Kate Tilleczek). In a review of qualitative research recruitment strategies for disadvantaged groups, Bonevski et al. (2014) concluded that members of these groups were difficult to find and providing a financial incentive could aid recruitment. Studies with NEET participants (Holte, 2017) and other marginalized youth (McCormick et al., 1999) reached similar conclusions. We also experienced challenges with recruitment of NEETs and decided to add a cash incentive of $10 (CDN/AUS) or £20 (UK) per participant as additional encouragement for this group of young people. All participants, including digital shadows, were then offered the incentive for each interview in which they participated, along with the Certificate.

Table 2.2 Actual Number of Original Participants

Category	In Education or Employment			NEET			Total
	Canada	Scotland	Australia	Canada	Scotland	Australia	
Third generation	30	11	22	6	4	0	73
Newcomer	13	0	1	0	0	0	14
Indigenous	13	0	0	4	0	1	18
	56	11	23	10	4	1	105

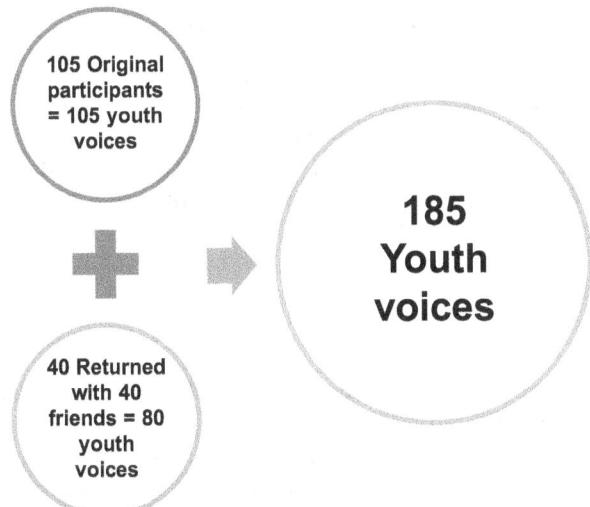

Figure 2.2 Voices Heard.

We engaged with 105 original youth participants (Table 2.2), 40 of whom returned for a second interview with a digital shadow. Those 40 were also participating in a dyadic format and brought new conversations to the table; thus a total of 185 youth voices were heard through interviews (Figure 2.2).

Phase 2 recruitment

At the time of writing, Phase 2 data collection was complete, but analysis had not begun. Decisions around selection of participants for Phase 2 were based on resources: time, money, and people; therefore, we elected to focus Phase 2 on Canada with a participation goal of 25% (n=17) of the 66 original

Canadian participants. Additionally, to ensure we had a representative sample, we included Indigenous participants from the Maritime provinces and all newcomer participants. We reached out to 52 young people who had been original Time 1 participants. If we had an email address, that was our first line of communication. Where we had no email address or did not receive a response via email, we used Facebook or Instagram. If we did not receive a response within a week, a second message was sent. If there was no response to the second message, the participant was not contacted again. At the end of Phase 2 data collection we had interviewed 21 of the original participants, 6 of whom brought a friend to the interview, for a total of 27 Phase 2 participants.

Developing ways in which to speak to youth

Youth recruited for Phase 1 were asked to participate in four ways: complete a short survey (Facesheet), participate in an interview and agree to be contacted for a second interview two years later, provide access to social media accounts, and recruit a friend (digital shadow) to participate with them in a second interview. In Phase 2, we eliminated the request for social media access and asked participants to engage in only one interview, with or without a friend.

Facesheet

A short demographic survey, which we called a Facesheet, was designed to elicit from participants basic demographic information, an overview of their online and social media activities, and to answer questions about benefits and problems with technology in their social and educational lives. For Phase 2, the Facesheet was simplified for ease of completion; other changes (e.g., the social media listed) were made based on analysis of Phase 1 data. One hundred twenty-one participants completed the Facesheet in Phase 1. All participants (*n*=27) in Phase 2 completed the Facesheet. These data were entered into SPSS 23 for descriptive analysis.

Interview

Participants were also asked to engage with researchers in a dynamic, semi-structured life history narrative interview and (Phase 1 only) agree to be contacted for a second interview two years later. Of the 105 original participants who were interviewed, 77 agreed to return for a second interview two years later. Digital shadows were not asked to return.

Although we were prepared to conduct interviews remotely using applications such as Skype, all Phase 1 interviews were conducted face-to-face by various members of the research team. Three Phase 2 participants were interviewed remotely via Skype (*n*=1) or telephone (*n*=2). The same master interview protocol was issued to the entire research team; however, each interviewer was encouraged to develop her or his own approach to the organization and wording of the interview questions so conversations with youth could flow as naturally as possible. The interview protocol was modified slightly for Phase 2 to focus on changes in social media use and impacts. Participants at all research sites were given the choice of interview location to maximize their comfort, including university meeting rooms, public libraries, community centres, youth training centres, and coffee shops.

Interviews began with the completion of the informed consent process; participants were provided with a written consent form outlining the project details and the ways in which they were being asked to participate. This document was also verbally explained by the interviewer. Once the consent form was reviewed, participants were asked to provide consent to each part of the process through a series of statements, each with a yes/no check box, and complete the process with a signature at the end of the document. Once informed consent was received, the interview commenced. Participants were asked to complete the survey, either on their own or by relaying their answers to the interviewer who then wrote them on the form.

Participant interviews were open and conversational in nature, though each interviewer would direct the conversation to project themes or questions, including exploring each participant's digital presence, their perspectives on digital media and its use, and their engagement with digital media for social and educational activities. The interviews also pilot-tested the efficacy of an "object task" where participants were provided with a computer or a cell phone and asked to "find yourself online". This task helped to initiate the interview discussion and allowed researchers to more directly experience a participant's online presence and persona(s). In some cases, this was not possible due to technological issues, primarily the lack of Internet access at the interview location. As another starting point for discussion, participants were also provided with an Icon Sheet which showed the icons, or logos, of the most common international digital applications and platforms (e.g., Facebook, Twitter, Instagram) and asked to indicate those which they used at the time of the interview, or had utilized in the past, and to add any others they used that were not represented on the Icon Sheet. Interviews typically lasted 30–90 minutes, depending on participant availability and interest. Interviews were audio and/or video recorded (or both) based on participants' preference and comfort level.

Social media

Participants were asked during the consent process to share some of their digital media activity with the research team. During the interview, interviewers explained how the data would be captured[3] and that the collection would last for three months following the interview. At the end of the interview this request was repeated to ensure that participants had ample opportunity to think about their response and ask questions about the process. The research team felt it was important to ask this again as the interview may have impacted the way participants felt about their privacy and security in their online life (see Chapter 4 for participant perspectives on these issues). If they agreed, they were then asked to provide their account name(s), or handle(s), so the Data Manager would be able to find the accounts and send a "Friend" or "Follow" request to gain access.

After the initial pilot interviews, researchers determined that it would be more efficient to have participants write their user names down themselves on a prepared form. The form initially asked youth to provide their handles (as applicable) for four social media applications (those revealed in initial discussions with youth to be the most popular at the time): Facebook, Twitter, Tumblr, and Pinterest. Many participants began manually adding Instagram on the form, so it was adjusted to include this as a fifth social networking site; a space was also added so youth could include access to any other platform they used regularly and wished to share with researchers. Participants were reminded that their participation was completely voluntary. Details of the social media collection process can be found in a YLRL white paper (Loebach & Madigan, 2015).

Of our original participants, 73 youth (70%) consented to participate in the social media capture. All of the digital shadows also agreed to participate in this study component, for a total of 113 youth.

Among all youth participants, 87 (60%) had provided their user name or handle on the form we provided. For the remainder, we searched each platform based on the information provided during the interview to locate and connect to their account. Several participants agreed that they would allow us to follow them on some platforms but not on others; often they used one platform for more public or "professional" communications, while others were shared with only close friends or family. Once permission was received, we began the social media data capture. Detailed collection notes were kept for each participant. These notes outlined the dates data were collected, the platform, and the format of the data (e.g., PDF, data set).

There were some instances where we were unable to confirm the identity of the participant within the social media platform, even with the information provided. Some participants did not respond to our friend or follow request, but we were able to capture publicly available information from those accounts. In some cases, the participant eventually accepted the request, allowing us to contrast which information was available to the public and which was kept private for friends or followers only. Several of the youth participants also changed their privacy settings or unfriended the research team during the three-month collection period; however, we had no communication from them to indicate why they wanted to change this arrangement.

Table 2.3 Number of Participants Who Shared per Type of Media

Facebook	Twitter	Instagram	Tumblr	Pinterest
80	26	26	8	5

In total, we were able to successfully capture social media data from 86 participants (54 original, 32 digital shadows) on one or more of their designated social media platforms for a period of three months. Some of the participants provided access to multiple platforms while others shared only one. The most common was Facebook, with 80 of the 86 participants using that platform. Table 2.3 indicates the number of participants who shared information from each of the five most common social media platforms.

Data from Facebook, Twitter, Tumblr, and Pinterest were captured in PDF format resulting in approximately 6,000 pages of data, much of it visual (i.e., photos rather than text). The Twitter data were in database (spreadsheet) format with approximately 77,000 lines of data captured.

Recruit a friend

Each original participant was asked to recruit a digital shadow and return with that person for a second, dyadic, interview with the research team. We did not provide any selection criteria for the shadow other than it should be someone with whom the participant engaged on a daily basis. Digital shadows were asked to complete the Facesheet and share some of their social media activity. They were not necessarily expected to remain in the study longitudinally. Forty digital shadows were recruited.

Documentary film

An independent filmmaker was contracted to develop a documentary film centred around our project. The documentary was conceived as a means of locating our project within the broader, more public, context of how the relationship between youth and digital media is understood. The filmmaker met with the Principal Investigator and Project Manager several times to discuss the project and his ideas; however, the final film was made at his discretion. Youth were selected from the existing pool of participants by project researchers in consultation with the filmmaker, who had elected to focus on pockets of resistance and the utopian/dystopian debate around digital media. Youth were contacted by members of the research team to determine their interest in being interviewed for the documentary; with their permission, contact information was forwarded to the filmmaker. At this point the filmmaker made direct contact with the participants, using his own protocols. The filmmaker determined that the best format for distributing the footage filmed was in a series of short video clips. Once completed, the clips were sent to the YLRL where opening and closing segments were added. The final videos can be seen on the Young Lives Research website (https://younglivesresearch.ca).

What were the ethical dilemmas considered/encountered?

The word "ethics" evokes the concept of doing the right thing. In research, ethics protocols focus on protecting participants from harm which includes protecting the data that are collected. We rely on formal research ethics guidelines to assist us in developing our research protocols and ensuring we are engaging with participants ethically. However, the guidelines available to us[4] did not reflect our methods, particularly in the qualitative collection of social media data. The literature on Internet research ethics is abundant but focussed primarily on methods and general guidelines (e.g., Convery & Cox, 2012; Eynon, Fry, & Schroeder, 2017; Hill et al., 2004; Markham & Buchanan, 2012). Research that used direct engagement with participants on social media was often through the use of Facebook groups (e.g., Bull et al., 2011; Lunnay, Borlagdan, McNaughton, & Ward, 2015). Given the paucity of literature to aid us in developing our ethical protocols, we took care to evaluate each data collection, transmission, storage, and analysis decision through an ethical lens and on an ongoing basis, engaging "Ethics in practice" defined by Warin (2011) as "a set of day-to-day practical negotiations and compromises" (p. 807).

Key guidelines in the TCPS2 (CIHR, NSERC, & SSHRC, 2014) focus on protecting the participants' identity throughout the whole of the project. To this end, no identifying, sensitive, or personal information associated with any participant was publicly shared or reported. Participants were assigned a unique ID code, which was used to label all documents, including digital data, related to that person, and care was taken to remove any identifying information from all project outputs, including identifying references (e.g., a participant's street or town name, or the name of their high school) within interview transcripts.

To ensure participants were further protected in dissemination activities, they were provided the opportunity to select a pseudonym or have one assigned (only two elected to choose their own). Therefore, all participant names used in dissemination are pseudonyms; care was taken to ensure that there was no duplication of names and that short forms of participant names were not used as pseudonyms.

There were exceptions in our project to the preservation of participant anonymity – specifically, in the use of dyadic interviews and video. Anonymity cannot be guaranteed when more than one participant is present (Bagnoli & Clark, 2010; Raby, 2010), or when disseminating visual data (Anderson & Muñoz Proto, 2016; Cox et al., 2014), such as the documentary film project and the digital portraits[5] which are described later in this chapter. These exceptions were clearly explained to the participants during the consent process. Additionally, at each point of the research process, youth were reminded that participation was voluntary, that they could choose whether to participate in each step, and that they could withdraw at any time.

Protecting the data

There were many different points in the research process when data security was particularly problematic. With a longitudinal and international study of this nature special attention must be given to the ethical collection, transfer, storage, and use of both physical (paper) and digital data. This project involved data collection at various sites within each of the three study countries (Canada, Scotland, Australia), posing substantial logistical issues for the ethical and efficient transfer of confidential research data to and from the YLRL, which was responsible for long-term storage.

The TCPS2 requires that any identifiable data sent over the Internet, or kept on a computer connected to the Internet, must be encrypted (CIHR, NSERC, & SSHRC, 2014, p. 63), and "computer passwords, firewalls, anti-virus software, encryption and other measures that protect data from unauthorized access, loss or modification" (p. 58) must be used. To supplement this, in the era of digital networking and increased online research, the Tri-Agency also created a "Statement of Principles on Digital Data Management" ([the Statement] CIHR, NSERC, & SSHRC, 2016) to further outline the ways in which researchers should store, access, and reuse digital data. A key requirement outlined in the Statement was the need for researchers to develop data management plans and ensure the guidelines of the TCPS2 were met. We worked through several processes to discover the best way to ethically transfer and store data resulting in the development of the YLRL Data Management Plan (see Figure 2.3). Central to this plan were a dedicated Data Manager and a secure (password-protected and not connected to the Internet) computer used only for data storage. Secondary storage was provided through a Virtual Research Environment (VRE) developed and maintained by the UPEI Robertson Library in consultation with the Data and Project Managers.

Social media data

The social media data collected presented a different set of ethical challenges. Although the data captured were saved on a password-protected external hard drive and transferred to the secure data computer, other ethical issues arose in the analysis stage. Most notably, the captured data included all activity on the

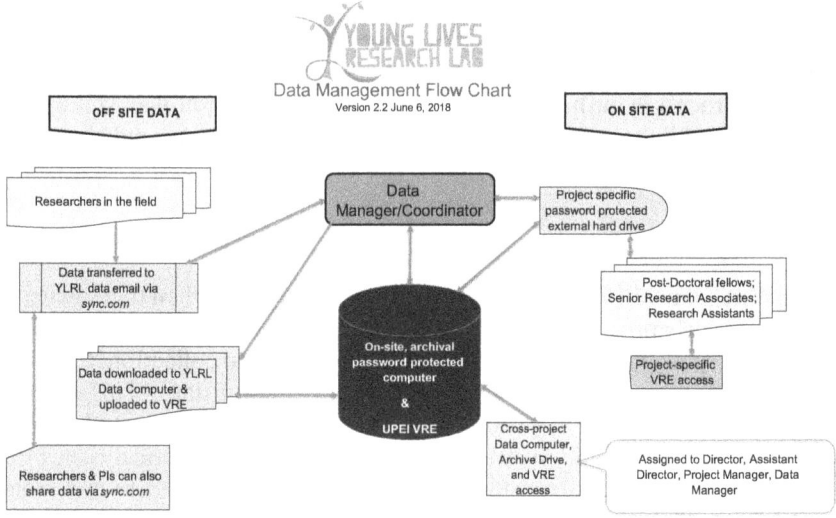

Figure 2.3 Data Management Plan.

account; therefore we had captured images and posts from non-participants. The volume of data collected was immense and time and resources did not allow for all non-participant data to be redacted. Researchers participating in the analysis of the social media data were required to exercise extreme caution in ensuring that they used only participant-generated material. Additionally, there would be no visual dissemination of this data with the exception of sections used in the digital portraits (where all data relating to anyone other than the participant was redacted).

Also of concern was the detailed metadata that was part of the capture process. This included user names; locations of the participant at the time of each posting; and, on platforms such as Twitter, information on non-participants (e.g., friends of the youth participant) who appeared in the captured data. Social media data were downloaded using NCapture, a tool within NVivo analysis software that allows for the capture of online content. It was not possible to modify NCapture to avoid acquiring this metadata; therefore, in instances where the captured data were to be shared with other members of the team for analysis, all such information was deleted, leaving only the content of the social media posts themselves.

At the end of Phase 1 data collection, the team assembled and prepared all data for analysis. All interviews were transcribed verbatim, either by members of the research team or through a contracted transcription service. Facesheets and Icon Sheets were scanned into PDF format. All social media communications and images from the three-month collection period were compiled chronologically in PDF format for each relevant platform. All documents and interview videos were uploaded to NVivo 10/11 software for coding and analysis.

In accordance with our Data Management Plan, all data were stored on a secure (password-protected, not connected to the Internet) computer and provided to researchers when requested using password-protected portable hard drives.

Analysis and interpretation

What do the data mean?

The project was conceived as a return to humanities-infused practices for youth; the investigative team was purposefully chosen for their cross-disciplinary scholarship and interpretations of young people and the modern digital age. The meanings we make from the data are grounded in these lenses and scholarship. Each chapter in this book and those to come illustrate the interdisciplinarity but youth-attuned interpretations we bring to bear on youth and the digital age. Data were coded into themes and narrative arcs (stories) of individual participants were developed and turned into digital portraits.

We employed trusted methods of reflexive analyses in three forms: thematic, narrative, and social organization (Alvesson & Sköldberg, 2018; Atkinson & Delmont, 2005; Braun & Clark, 2006; Bruner, 2004; Daiute, 2013) from verbatim transcripts and visual texts. Observation, interviewing, and textual analysis were triangulated and checked back against the daily practices of youth (Smith, 2002). Analysis provided theoretical insights with a sufficient degree of transferability to comparable contexts. Rigour in field work and analysis relates to the trust-worthiness of findings addressed in critical ethnography by asking if it does indeed look and work this way (Lincoln, Lynham, & Guba, 2011). Rigour was carefully considered in a manner appropriate to qualitative research (Fossey, Harvey, McDermott, & Davidson, 2002; Twohig & Putnam, 2002) with attentiveness to research practice through reflection, sensitivity to context, detailed audit trails of processes, and team analysis. This book presents initial themes which emerged through the early stages of analysis.

Emerging themes

We began by reviewing and coding interview data to identify emerging themes, also considering themes that had been identified by interviewers during data collection and shared with team members. For example, one issue that was raised by many participants early in the study was a concern for the younger generation; "I think social media in general is just kind of messing up the generations coming up. I don't know why. Just nobody's, social networking is just, nobody actually talks to people in person" (Owen, 18)[6]. This particular theme was unexpected as it did not appear in our reviews of existing literature. It was, however, a concern shared almost universally by our participants. When this became apparent, the team considered adding a question about this concern for younger people into the protocol but elected to let it continue to emerge organically. These early-emergent themes were considered in the ongoing analysis; transcripts were revisited to verify whether elements of any identified theme were present and were coded accordingly. Data were reviewed by several researchers and results compared to check inter-coder reliability. Researchers engaged in ongoing, informal discussions as necessary and met weekly (bringing in off-site researchers via Skype) to discuss and compare findings.

Digital portraits

The breadth and variety of the data called for the incorporation of a variety of analytic methods and types of dissemination. Video recording of the interviews provided an opportunity for narrative analysis in the form of *digital* (video and narrative) *portraits* of participants. This process was piloted in 2014–2015 with seven cases – selected based on the completeness of the data collection and recommendations from interviewers of partici-pants found to be responsive and dynamic in interview. Data collected for each participant were reviewed and notes made for the portrait. The pilot resulted in the creation of one complete narrative and video digital portrait. Notes were prepared for the remaining six cases. In 2017, the notes for the pilot cases were retrieved and expanded, and another case was added at the request of the PI. Two types of digital portrait were prepared: (1) one-page written narrative and (2) three-minute video.

Each participant interview contained a wealth of data and demonstrated several potential narrative arcs. However, early analysis indicated a recur-ring theme of tensions surrounding the relationship of youth with/to social media, in particular expressions of media as both good and bad and con-cern for the future of technology in their lives. Therefore, those were the themes selected for the first set of digital portraits.

The written narrative provided a descriptive analysis of the participant based on the data in the summary notes. It was designed to be read like a story, beginning with demographics, adding some text about the participant's beliefs about digital media and its impact on social and educational life, and ending with a "wrap-up" or "final thoughts" statement. Direct quotes were used to illustrate each main point.

Following the narrative, the producer (a member of the research team) reviewed the video and highlighted sections on the transcript that fit the selected themes. These sections were timestamped and given to a film editor (engaged specifically for this purpose) as a script to follow. The editor then cut the original video to match the script. The shortened video was reviewed by the producer and any changes to the script were noted and returned to the editor for further adjustment. This process continued until the producer and editor agreed that the finalized video was coherent and reflected the character of the participant. Where necessary, title slides were placed in the video to reflect a change of topic. Finally, the editor created a music score as well as opening and closing credits. The result was a three-minute digital portrait.

The digital portraits were designed to be used as exemplars during presentations about this work to showcase what our participants were telling us, using their own voices, in "lived time" (Bruner, 2004, p. 692). There were eight portraits created and forwarded to participants for their approval and permission to use for academic purposes. Three participants did not respond to the request for permission. Where permission was granted, the digital portraits were uploaded to the YLRL YouTube page as unlisted (see Chapter 1 for links), meaning that only those who were provided the URL would be able to view them, and they would not be found in a search. The permission to share the videos was based on our use of them in academic presentations; therefore they are not public or searchable on YouTube, nor are they posted to the YLRL website.

Methodological challenges in this study

The length, depth, and breadth of this project allowed for many opportunities to evaluate methods. As questions arose, we discussed them as a team to determine best practices. Some of the areas that required consideration were recruiting young people, managing multiple interviewers/sites, and ongoing consent.

Recruiting young people

Finding new and successful ways to recruit youth participants can be challenging (Amon, Campbell, Hawke, & Steinbeck, 2014; Henderson, Law, Palermo, & Eccleston, 2012; Loxton et al., 2015). We implemented a variety of recruitment methods with varied results. Although some youth heard about the study through recruitment posters and presentations and subsequently contacted the research team independently, many youth participants were drawn in to the study by a local youth leader. This leader would be someone within the youth-related organization who took interest in the study and took it upon themselves to recruit youth to participate. The leader often acted as a liaison between the youth and the research team, helping to schedule interviews, encouraging youth commitment to the project, and providing space to conduct interviews. Working with these youth leaders and/or making an in-person presentation to a youth group proved to be the most effective method for recruiting young people to the project and a lesson to take forward for future projects.

When endeavouring to contact interested youth to set up interview schedules, we found that many young people did not often use a phone conventionally and many did not have an email address they used for communication. For some, online communication, such as texting, Facebook messaging, or "IMing" (Instant Messaging) through other social networking sites, is more common and often preferred as the primary mode of contact. It was important to engage youth using their media of choice, as phone messages and emails would often go unanswered. Requests communicated through social media tended to solicit faster responses, and, as youth would often confirm or alter scheduling details at the last minute, these more mobile modes of communication appeared more compatible with the schedules and practices of contemporary youth. It is also possible that the social media accounts of youth are more stable than cell numbers or email addresses; therefore, we shifted our communication protocol before we approached youth for Phase 2 to include cell phone, texting, or Facebook messaging to reach or stay in contact with youth participants.

Managing multiple interviewers/sites

To ensure that we could reach our participation goals, and due to the geographical distance between research sites, we engaged the services of several interviewers outside the YLRL. We developed an interview guide to be shared with all researchers to ensure everyone was following the same protocols. At the same time, it was necessary that there be flexibility within the protocols to allow for adjustments based on local circumstances. During the analysis, several issues around interviewing techniques were noted. For example, some interviewers focussed on asking all the questions on the interview protocol while others deviated in order to follow up on certain participant statements. There were also incidents of leading the direction of the conversation with statements such as "clearly it's [social media] important in your life" rather than framing it as a question. Transcription of interviews was done in batches through an external service and there was no process in place to review interviews as they were received. It is important to establish a pattern of reviewing early interviews immediately after they are received to ensure that interviewers are both following the protocol and being responsive to what participants are saying.

Managing remote interview sites can be challenging. Most of our communication with interviewers in Northern Ontario, Scotland, and Australia was via email and often interviewers were slow to respond. Our processes for receiving the data from remote sites were complex and evolving, resulting in delays. It would have been more productive to have interviewers transmit material immediately following each interview (or at the end of each day) rather than in batches. This would have ensured the data were received in a timely manner and any issues could be dealt with promptly.

Ongoing consent

Our commitment to ethical research is strong and we made every effort to ensure that participants knew what we were asking of them and what we planned to do with their data, practising "ethical mindfulness" (Warin, 2011). This sometimes took the form of asking for the same permissions multiple times; for example, asking permission to collect social media data during the consent process and again at the end of the interview. Participants may have reconsidered as the interview progressed and they began to think more about the role of social media in their lives. Checking with participants at each stage of the research or "process consent" (Ellis, 2007) is critical to ensuring that they have not changed their mind or become disengaged from the project. Sarah Pink (quoted in Cox et al., 2014, p. 13) reflected on her methods as

> having a staged or layered process of consent, which is individual to each participant depending on what they agree to and which stages they wish to be consulted at. Therefore I might, depending on what has been agreed, send participants copies of images to approve before they are archived, published, or shown, should they wish me to do so. Sometimes consent is revisited after a period of time. I believe that creating processes of ongoing consent offers a useful way of doing this, even though it can create a lot more administration for the researcher. It ensures that if participants want it, lines of contact can be maintained.

Like Pink, we also endeavoured to send copies of the videos to participants for their review and approval. However, we encountered difficulties connecting with participants after the completion of the digital portraits, and, as a result, we were unable to secure final approval of three of the portraits. In Phase 2, we used a separate consent form for dissemination after providing participants with a sample clip from their first interview to give them a clearer understanding of what we were asking. That consent form included use (for academic purposes) of selected clips from both Phase 1 and Phase 2 interviews. Vetting of the portrait and further permission would not be necessary.

Conclusions

We began this chapter with Cohen and Ainley's (2000) call for new methodological approaches to youth research and with a will to countenance methodological anarchy as per Feyerabend (1993) and Mills (1959). This call does not eliminate method but rather takes notice of method that is open and flexible, and focussed on context of a specific project rather than adherence to a set of discipline-specific rules. We engaged our sociological imagination to gain understanding of social media in young lives through the lens of its presence and place in society as a whole (Mills, 1959, p. 3). We had a planned process for data collection and analysis, a decision-making process, and a set of research tools. Our methodological openness enabled us to continuously evaluate our progress and goals, making adjustments as necessary to do justice to an authentic project and the young people we were representing while maintaining our commitment to ethical practice. It was a highly reflective and youth-attuned process.

We employed a youth-attuned qualitative design (Tilleczek, 2018a) which included audio- and/or video-recorded face-to-face interviews with 105 youth participants, an invited dyadic interview with their friends (digital shadows; $n=40$), a demographic survey, and the capture and analysis of images and text from participants' digital media activity on platforms such as Facebook, Instagram, Twitter, Pinterest, and Tumblr. Whenever we identified an issue, we engaged in rigorous discussion and debate as to the direction we should take, as in the question of altering the interview protocol to include the emerging theme of concern for younger generations. We revised our recruitment strategies for NEETs and digital shadows when our original plans were not effective. We added a process to obtain social media information from participants as a means of both locating their media and ongoing consent. Finally, Phase 2 research tools were modified to reflect what we learned during our analysis of Phase 1. All of these adaptations were in direct response to our continuous reflection on our methods.

Technology is changing at an unprecedented pace. In the course of this project, young people have moved from Facebook to Instagram and Snapchat. Every day there is something new. Issues of surveillance and privacy are front and centre in 2018 much more than when we began speaking to youth in 2014. So, although our participants spoke of this, the scale and impact today is significantly different than it was at the time of the initial interviews. We anticipate more in-depth discussion of this in our subsequent analyses and books.

Our participants were chosen because they represented a unique moment in time; they were not born into a technological world but grew up in it. Many expressed concerns for the next generation; replicating this study with current 16–19 year-old-participants would provide the opportunity to determine if those concerns were valid.

We began with the premise that methods are important but must be flexible and responsive to the phenomenon of study. We demonstrated this fluidity throughout our project, adapting our approach as required through serious discussion of events and issues and responsiveness to participant needs. In doing so, we developed strategies and techniques, some of which became the basis of white papers which have been posted to our website (https://younglivesresearch.ca).

We closely follow the work of others in the field, such as *MediaSmarts* (http://mediasmarts.ca/) and *Global Kids Online* (http://globalkidsonline. net/). MediaSmarts has produced a large body of data about Canadian youth online; much of it related to their quantitative survey "Young Canadians in a Wired World". They primarily disseminate their data through reports and infographics. Global Kids Online is a longitudinal, international partnership focussed on children's use of Internet around the world. In addition to research reports, Global Kids Online has developed a set of research tools and method guides that can be downloaded and shared. The *Digital Media and Young Lives* project is similar to the work that these organizations do in the focus on youth and technology. However, our youth-attuned qualitative methods provide a breadth of data beyond what is produced elsewhere. For example, the digital portraits created in this project provide a unique opportunity to hear young people speak in their own voices and to multiple audiences.

Young people can hold a triply colonized position from which to enact their agency and resistance in the digital age. Given the enormous pall of technology and digital media on the lives of youth, the ambition for big technology companies to colonize the spaces of young lives for profit, and the scant space that youth have had to discuss and debate these influences, we approach this work from the perspective of qualitative youth research as decolonization and decolonality (Tilleczek, 2018b). We envision processes of decolonization as striving to "apportion power and knowledge from the relatively privileged (i.e., adults, medical and psychological sciences, and the global north) to the relatively less privileged (i.e., youth, social science, humanities and the global south)" (Tilleczek, 2018b). Using Maldonado-Torres's (2016) concept of decolonality, we further strive to privilege the voices of our youth participants and envision our work as a "collective project" (p. 28) wherein the voices of the colonized take precedence.

Our methodological approach was developed in line with our understanding of decolonizing youth studies through research processes *with, for,* and *by* youth (Tilleczek, 2018b). The researcher-participant relationship is one of perceived power, particularly when the researcher is an adult and the participant a youth. We worked to ameliorate that power dynamic through our methods, such as having the participant recruit their own digital shadow and select the social media platforms they would share. Additionally, our commitment to ongoing consent ensured that they were continually engaged in the decision-making process. And, our open and conversational interview processes were meant to provide a space for youth to discuss these issues of importance to them, a space that they assured us is not common for them.

It is our hope that our ways of seeing can help to create a new foundation for speaking with young people within and about the modern digital age. We fully expected that the young people with whom we spoke would help us to refine our interview protocols and develop unique and compelling video footage, digital data, and documentary film analysis in order to understand and communicate our findings. Already the methods we have employed have yielded volumes of fascinating narrative, visual, video, and social media data from which we are able to illustrate the varied and extraordinary stories, experiences, and practices of these young people. The analysis illustrates the logic and illogic of the digital age and asks youth to reflect on their place within it.

We have begun to see points of resistance and reproduction occurring (see Tilleczek, 2017a; Tilleczek & Tilleczek, 2016). On the one hand, many of the youth live a technologically ubiquitous and cherished form of life. On the other hand, many also see in this form of life something that troubles them because of what it is costing them emotionally, socially, and intellectually and how it is affecting "the little kids" in their lives. Many have told us that they can see that screen time is eating up the real time, energy, concentration, and human connection necessary for young children to grow. But they also feel that "they can handle the problems" when it comes to their own lives. We asked them why and how this all holds together for them. The individual and dyadic interviews reveal a deep desire for more spaces in which to discuss and debate the digital age and for youth-led responses and critical actions. Our study opened up one such space.

Notes

1   The authors would like to thank Jonah Rimer, Matthew Munro, and Heather Barnick for their input on data analysis and reviews of this chapter. We would also like to acknowledge Ruby Madigan for her earlier contributions to data management.
2   Team members included: Principal Investigator (Dr Kate Tilleczek), Co-Investigators (Dr Andy Furlong, Dr John Smyth, Dr Ron Srigley, and Dr Katherine Boydell), Project Manager (Valerie Campbell).
3   The program used to gather social media data was called "NCapture"; thus we use "capture" to identify the way in which this portion of the data was collected.
4   In 2001, three major research agencies in Canada, the Social Sciences and Humanities Research Council of Canada (SSHRC), the Natural Sciences and Engineering Council of Canada (NSERC), and the Canadian Institutes of Health Research (CIHR), known collectively as the Tri-Agency, created the Interagency Advisory Panel on Research Ethics (PRE). The PRE (http://www.pre.ethics. gc.ca) is responsible for development, interpretation, and implementation of the *Tri-Council Policy Statement: Ethical Conduct for Research Involving Humans* (TCPS2). It is this policy that guides our institutional Research Review Boards (REBs), which, in turn, provide guidance to our research.
5   Short (three-minute) narrative portraits developed from interview transcripts and videos, Facesheets, and social media data. Further described in Analysis section.
6   All participants have been given a pseudonym.

References

Alvesson, M., & Sköldberg, K. (2018). On reflexive interpretation: The play of interpretive levels. In M. Alvesson & K. Sköldberg (Eds.), *Reflexive methodology: New vistas for qualitative research* (3rd ed., pp. 321–342). London, UK: Sage.

Amon, K. L., Campbell, A. J., Hawke, C., & Steinbeck, K. (2014). Facebook as a recruitment tool for adolescent health research: A systematic review. *Academic Pediatrics, 14*(5), 439–447.

Anderson, S. M., & Muñoz Proto, C. (2016). Ethical requirements and responsibilities in video methodologies: Considering confidentiality and representation in social justice research. *Social and Personality Psychology Compass, 10*(7), 377–389.

Atkinson, P., & Delamont, S. (2005). Analytic perspectives. In N. K. Denzin & Y. S. Lincoln (Eds.), *Handbook of qualitative research.* (3rd ed., pp. 821–840). Thousand Oaks, CA: Sage.

Bagnoli, A., & Clark, A. (2010). Focus groups with young people: A participatory approach to research planning. *Journal of Youth Studies, 13*(1), 101–119.

Baker, S. E., & Edwards, R. (2012). *How many qualitative interviews is enough? Expert voices and early career reflections on sampling and cases in qualitative research* (Review Paper No. NCRM/019). Southampton, UK: National Centre for Research Methods.

Bell, B. L., & Campbell, V. M. (2014). *Dyadic interviews in qualitative research.* Charlottetown, CA: Young Lives Research Laboratory. Retrieved from http://katetilleczek. ca/files/2014/08/RS1-Dyadic-Interviews-in-qualitative-research-Nov-2014.pdf

Bonevski, B., Randell, M., Paul, C., Chapman, K., Twyman, L., Bryant, J., … Hughes, C. (2014). Reaching the hard-to-reach: A systematic review of strategies for improving health and medical research with socially disadvantaged groups. *BMC Medical Research Methodology, 14*(1), 42.

boyd, d. (2007). Why youth (heart) social network sites: The role of networked publics in teenage social life. In D. Buckingham (Ed.), *Youth, identity, and digital media* (pp. 119–142). Cambridge, MA: MIT Press.

Braun, V., & Clarke, V. (2006). Using thematic analysis in psychology. *Qualitative Research in Psychology, 3*(2), 77–101.

Bruner, J. (2004). Life as narrative. *Social Research, 71*(3), 691–710.

Bull, S. S., Breslin, L. T., Wright, E. E., Black, S. R., Levine, D., & Santelli, J. S. (2011). Case study: An ethics case study of HIV prevention research on Facebook: The just/us study. *Journal of Pediatric Psychology, 36*(10), 1082–1092. doi:10.1093/jpepsy/jsq126

Canadian Institutes of Health Research, (CIHR), Natural Sciences & Engineering Research Council of Canada (NSERC), & Social Sciences & Humanities Research Council of Canada (SSHRC). (2014). *Tri-council policy statement: Ethical conduct for research involving humans 2014* (Report No. RR4-2/2014E). Ottawa, ON: Interagency Secretariat on Research Ethics. Retrieved from http://www.pre. ethics.gc.ca/pdf/eng/tcps2-2014/TCPS_2_FINAL_Web.pdf

Canadian Institutes of Health Research, (CIHR), Natural Sciences & Engineering Research Council (NSERC), & Social Sciences & Humanities Research Council (SSHRC). (2016). *Tri-agency statement of principles on digital data management.* Retrieved from http://www.science.gc.ca/eic/site/063.nsf/eng/h_83F7624E.html? OpenDocument

Cohen, P., & Ainley, P. (2000). In the country of the blind? Youth studies and cultural studies in Britain. *Journal of Youth Studies, 3*(1), 79–95. doi:10.1080/136762600113059

Convery, I., & Cox, D. (2012). A review of research ethics in internet-based research. *Practitioner Research in Higher Education, 6*(1), 50–57.

Cox, S., Drew, S., Guillemin, M., Howell, C., Warr, D., & Waycott, J. (2014). *Guidelines for ethical visual research methods.* Melbourne, VIC: University of Melbourne.

Daiute, C. (2013). *Narrative inquiry: A dynamic approach*. Thousand Oaks, CA: Sage.

Dworkin, S. (2012). Sample size policy for qualitative studies using in-depth interviews. *Archives of Sexual Behavior, 41*(6), 1319–1320. doi:10.1007/s10508-012-0016-6

Ellis, C. (2007). Telling secrets, revealing lives: Relational ethics in research with intimate others. *Qualitative Inquiry, 13*(1), 3–29.

Eynon, R., Fry, J., & Schroeder, R. (2017). The ethics of online research. In N. G. Fielding, R. L. Lee, & G. Blank (Eds.), *The SAGE handbook of online research methods* (2nd ed., pp. 19–37). Thousand Oaks, CA: Sage.

Feyerabend, P. (1993). *Against method* (3rd ed.). New York, NY: Verso.

Fossey, E., Harvey, C., McDermott, F., & Davidson, L. (2002). Understanding and evaluating qualitative research. *Australian & New Zealand Journal of Psychiatry, 36*(6), 717–732. doi:10.1046/j.1440–1614.2002.01100.x

Furlong, A. (2006). Not a very NEET solution: Representing problematic labour market transitions among early school-leavers. *Work, Employment & Society, 20*(3), 553–569. doi:10.1177/0950017006067001

Henderson, E. M., Law, E. F., Palermo, T. M., & Eccleston, C. (2012). Case study: Ethical guidance for pediatric e-health research using examples from pain research with adolescents. *Journal of Pediatric Psychology, 37*(10), 1116–1126.

Hennink, M. M., Kaiser, B. N., & Marconi, V. C. (2017). Code saturation versus meaning saturation. *Qualitative Health Research, 27*(4), 591–608. doi:10.1177/1049732316665344

Hill, M. L., King, C. B., Eckert-Denver, C., Gibson, E., Pankoff, B., & Rice, T. (2004, May). *The ethics of online research: Issues, guidelines and practical solutions*. Paper presented at the Society for Prevention Research Conference, Quebec City, QC.

Holte, B. H. (2017). Counting and meeting NEET young people: Methodology, perspective and meaning in research on marginalized youth. *Young, 26*(1), 1–16.

James, C., Davis, K., Flores, A., Francis, J. M., Pettingill, L., Rundle, M., & Gardner, H. (2010). Young people, ethics, and the new digital media. *Contemporary Readings in Law and Social Justice, 2*(2), 215–284.

Lincoln, Y. S., Lynham, S. A., & Guba, E. G. (2011). Paradigmatic controversies, contradictions, and emerging confluences, revisited. In N. K. Denzin, & Y. S. Lincoln (Eds.), *The SAGE handbook of qualitative research* (4th ed., pp. 97–128). Thousand Oaks, CA: Sage.

Livingstone, S., & Brake, D. R. (2010). On the rapid rise of social networking sites: New findings and policy implications. *Children & Society, 24*(1), 75–83.

Loebach, J., & Madigan, R. (2015). *Collecting social media data for qualitative research*. Charlottetown, ON: Young Lives Research Laboratory. Retrieved from http://katetilleczek.ca/files/2014/08/Collecting-Social-Media-Data-for-Qualitative-Research.pdf

Loxton, D., Powers, J., Anderson, A. E., Townsend, N., Harris, M. L., Tuckerman, R., … Byles, J. (2015). Online and offline recruitment of young women for a longitudinal health survey: Findings from the Australian longitudinal study on women's health 1989–95 cohort. *Journal of Medical Internet Research, 17*(5), e109. Retrieved from http://www.ncbi.nlm.nih.gov/pubmed/25940876

Lunnay, B., Borlagdan, J., McNaughton, D., & Ward, P. (2015). Ethical use of social media to facilitate qualitative research. *Qualitative Health Research, 25*(1), 99–109.

Maldonado-Torres, N. (2016). *Outline of ten theses on coloniality and decoloniality.* Foundation Frantz Fanon, Retrieved from http://frantzfanonfoundation-fondationfrantzfanon.com/IMG/pdf/maldonado-torres_outline_of_ten_theses-10.23.16_.pdf

Markham, A., & Buchanan, E. (2012). *Ethical decision-making and internet research: Recommendations from the AoIR ethics working committee (version 2.0).* Association of Internet Researchers. Retrieved from https://aoir.org/reports/ethics2.pdf

McCormick, L. K., Crawford, M., Anderson, R. H., Gittelsohn, J., Kingsley, B., & Upson, D. (1999). Recruiting adolescents into qualitative tobacco research studies: Experiences and lessons learned. *Journal of School Health, 69*(3), 95–99.

McCracken, G. (1988). *The long interview.* Newbury Park, CA: Sage.

Mills, C. W. (1959). *The sociological imagination.* New York, NY: Oxford University Press.

Morgan, D. L., Ataie, J., Carder, P., & Hoffman, K. (2013). Introducing dyadic interviews as a method for collecting qualitative data. *Qualitative Health Research, 23*(9), 1276–1284. doi:10.1177/1049732313501889

Patton, M. (1990). *Qualitative evaluations and research methods* (2nd ed.). Newbury Park, CA: Sage.

Raby, R. (2010). Public selves, inequality, and interruptions: The creation of meaning in focus groups with teens. *International Journal of Qualitative Methods, 9*(1), 1–15.

Smith, D. E. (2002). Institutional ethnography. In T. May (Ed.), *Qualitative research in action* (pp. 17–52). London, UK: Sage.

Smyth, J., & Harrison, T. (2015). The 'Hidden transcripts' of digital natives in the peri-urban jungle: Young people making sense of their use of social/digital media. *Educational Practice and Theory, 37*(1), 5–17. doi:10.7459/ept/37.1.02

Srigley, R., & Tilleczek, K. (2016, Jul). *Youth and the digital age.* Paper presented at the International Sociological Association World Forum, Vienna, Austria.

Tilleczek, K. (Ed.). (2008). *Why do students drop out of high school? Narrative studies and social critiques.* Lewiston, NY: Edwin Mellen Press.

Tilleczek, K. (2017a, Oct). *Resistance is futile! Young cyborgs & warriors in the digital age.* Keynote address for the session Child & Youth Engagement: Civic Literacies and Digital Ecologies at the Social Justice Research Centre and Department of Child and Youth Studies, Brock University, St Catherines, ON.

Tilleczek, K. (2017b, Dec). *The digital and age youth wellbeing: Digital media and young lives over time and place.* Invited keynote panel for Canadian Institutes for Advanced Research (CIFAR) workshop on Adolescent development and wellbeing in low and middle-income countries (LMICS), University of Oxford, Oxford, UK.

Tilleczek, K. (2018a, Jul). *Digital capital in late modernity: Andy Furlong's legacy.* Invited keynote for The Legacy of Andy Furlong Panel: Sociology of Youth (RC34) at the World Congress of the International Sociological Association, Toronto, ON.

Tilleczek, K. (2018b). Qualitative methodology in adolescent research. In S. Hupp & J. Jewell (Eds.), *The encyclopaedia of child and adolescent development* (Volume 1). New York, NY: Wiley.

Tilleczek, K. C. (forthcoming) *Young cyborgs? Youth resistance to technology.* Toronto, ON: University of Toronto Press.

Tilleczek, K., & Srigley, R. (2017). Young cyborgs? Youth in the digital age. In A. Furlong (Ed.), *Routledge handbook of youth and young adulthood* (2nd ed., pp. 273–284). New York, NY: Routledge.

Tilleczek, K., & Tilleczek, E. (2016, Jul). *Young cyborgs: Rituals of resistance to technology.* Paper presented at the International Sociological Association World Forum, Vienna, Austria.

Twohig, P. L., & Putnam, W. (2002). Group interviews in primary care research: Advancing the state of the art or ritualized research. *Family Practice, 19*(3), 278–284.

Warin, J. (2011). Ethical mindfulness and reflexivity: Managing a research relationship with children and young people in a 14-year qualitative longitudinal research (QLR) study. *Qualitative Inquiry, 17*(9), 805–814.

Chapter 3

Youth well-being and digital media

Kate C. Tilleczek, Brandi L. Bell,
and Matthew Munro

Youth well-being

A team of global experts at a recent Canadian Institute for Advanced Research (CIFAR)-funded workshop came together in Oxford to discuss new models and measures of youth well-being that could bear on identified research gaps (Young Lives, 2017). It was agreed that comprehensive models of youth well-being must be developed in situ and explore the range of impacts of digital technologies on youth well-being, including digital technology access patterns, forms of usage, risks, and benefits, particularly in ways that are attuned to culture and context. Models of youth well-being should weave biological, environmental, social, and economic measures and evidence together to provide transformative understandings. It was also agreed that ongoing evidence is required about how the Internet and digital technologies could be leveraged to promote well-being and revolutionize supports, while avoiding deleterious consequences.

Well-being has been conceptualized and measured in various ways. In Canada, initiatives are underway that attempt to model and measure well-being with fulsome measures that are in stark contrast to singular economic measures such as gross domestic and/or national product (GDP/ GNP). The Canadian Index of Wellbeing (CIW) is the best known set of such concepts and measures with a focus on Canadian adults (Michalos et al., 2011). The scholars who developed the CIW define well-being in the broadest sense and attempt to shift the conversation to a social determinants model. They define well-being as "the highest possible quality of life in its full breadth of expression" that includes good living standards, sustainable environments, robust health, vital communities, an educated population, balanced time use, high levels of democratic participation, and access/participation to leisure pursuits (Canadian Index of Wellbeing, 2016, p. 11). They measure these lofty societal and individual outcomes across eight domains with eight indicators for each using a total of 64 indicators for the CIW. This research and evaluation framework:

> Examines, tracks, and reports on how people are really doing in respect to the broad determinants of health … [and] compiles quantitative data

on eight interconnected quality of life domains that Canadians really care about; Community Vitality, Democratic Engagement, Education, the Environment, Healthy Populations, Leisure and Culture, Living Standards, and Time Use. (Association of Ontario Health Centres, n.d., para. 2)

The architects of the CIW and the United Nations International Children's Emergency Fund (UNICEF)'s Index of Child Well-being (Azzopardi, Kennedy, & Patton, 2017) are among those who are now at work with UNICEF Canada to develop a Child/Youth Well-being Index for Canada (CY-Index, UNICEF Canada, n.d.). This process began in 2016 and continues into 2019 with four phases of development and pilot testing. Pertinent to youth well-being in the digital age, the process, model, and resultant CY-Index suggest where and how we can begin to develop a nuanced understanding of digital media in young lives. It is based also on the fact that Canada has scored 24th out of 29 countries on youth rating of their sense of well-being and 17th out of 29 countries on the overall measure of well-being as part of the UNICEF Index of Child Well-being for affluent countries (UNICEF Canada & Students Commission of Canada, 2017). Canada also ranked 25th out of 41 countries on the Index of Child and Youth Well-being and Sustainability (UNICEF Canada, 2017). In the wake of persistently low rankings of child and youth well-being among high-income countries, UNICEF Canada is endeavouring to understand how this problem has become entrenched. Why are Canadian youth scoring so low? What is happening in their lives that is compromising well-being? What is it about Canadian society that negates well-being for youth?

While many social, cultural, and individual factors are at play in these trends, it is odd that neither the UNICEF Index nor the CY-Index have so far reckoned with the pervasive encroachment and ubiquity of digital media into young lives. The 2017 Report *My cat makes me happy: What children and youth say about well-being* (UNICEF Canada & Students Commission of Canada, 2017) illustrates that models of well-being could be more inclusive of the following domains as suggested by young people: health, relatedness, equity, education and employment, youth engagement, affordable living conditions, and space and environment. The report outlines the domains and associated themes as follows:

- Health: mental, sexual/reproductive, physical, and spiritual health in additional to access to appropriate healthcare
- Relatedness: experiences of belonging, having stable relationships with caring adults, friendship, love, online community, and social interaction
- Equity: discrimination, diversity, and respect

- Education and employment: inclusive education, employment, and future opportunities
- Youth engagement: youth activities, participation in decision-making, youth spaces, and youth contributions
- Affordable living conditions: food, housing, transportation, material resources, economic opportunity, and public resources
- Space and environment: nature/environment, positive living conditions, and supportive community

The report describes the important work accomplished to bring youth perspectives to the concept and measurement of well-being yet does not clearly explicate the place and role of digital media and technology, although there are some exceptions: the affordable living conditions domain is envisioned to include access to Wi-Fi; the relatedness domain encompasses building online community, specifically monitoring "cyberbullying, access to Internet, and social media" (p. 9); the youth engagement domain includes accessible online spaces and inclusive media; and the equity domain includes the need for diverse media representations. While these are important considerations, we argue that the conceptualization of youth well-being must better integrate access, skill, risk, opportunity, and use of digital media within and beyond the current domains.

Youth perspectives on digital media and well-being

With the aforementioned considerations about well-being in place, we turn to stories from youth who shed light on where and how to better understand the influences of digital media on well-being. Chapter 2 details the methodological and ethical decisions that were taken to conduct the research and guide analysis. The data presented here are from 66 individual interviews with Canadian youth from the study. This youth sample ranges in age from 15 to 26 years and is equally divided between female and male. Each interview transcript was reviewed and coded purposefully for domains of well-being from the UNICEF model, in addition to other factors of well-being that youth discussed but did not necessarily fit into the existing domains. Interview transcripts were reviewed to determine how young people spoke about the identified domains of well-being in the context of their use of digital media. The following questions guided the analysis: How does digital media use figure into each domain? What considerations are being shared by young people for understanding well-being in the context of the digital age? Do the current domains encompass the aspects of digital media that youth are speaking about? Is there anything missing? We outline the most salient and representative of the youth responses next, organized by domain.

Health

The youth provided deep insight into the health domain as relates to digital media. However, their responses mainly covered themes of mental health and physical health with relatively less discussion of spiritual or reproductive health. As relates to mental health, the constant pressure to keep up with new devices and to engage with social media platforms was generally described as a negative influence. Participants described the addictive qualities of digital media, saying that it was difficult to unplug and turn off: "I believe that this is the most dangerous addiction. It is because you can't, you don't even know that you're addicted to it" (Erika,[1] 19). Over consumption of digital media, distraction from other tasks, and the social expectation to constantly check messages also left many youth participants feeling anxious:

> I think that's one of the unfortunate things about the digital era is that some people get into it and they just can't leave, they can't bear the thought of not knowing what's happening to someone or what's happening over there, they just can't bear with it, they need it, almost like an addiction, for some people. (Elijah, 19)

Online violence was a part of social life for many youth participants and digital media was not blamed but rather seen as a new medium wherein hurtful images, texts, and drama can spread extremely quickly. Content that started out as innocuous could end up becoming very hurtful. There was a belief that those who already have mental health issues, particularly anxiety and depression, would gravitate to social media to share their experiences, which in some cases resulted in peer support and in other cases resulted in drama and violence. Some participants discussed how digital media has had a negative impact on their physical health, explaining that online gaming and time spent on social media have reduced their physical activity:

> If I didn't have access to the Internet, I'd get so much more done. I'd probably be in great shape, I'd eat better, probably, I'd have more time to do things that are productive instead of wasting time watching things and learning about gaming stuff that doesn't matter. (Easton, 17)

However, others provided examples of how they actively managed their time spent on social media to reduce their digital footprint and improve their physical health:

> But since I've gotten more active I've had a much smaller ... digital footprint. At the same time though, I feel more fulfilled ... for the first time ever in my life, because I'm doing so much extra work, because I'm actually improving my body I have more energy. Like, the physical feels so much better. (Cooper, 19)

Online resources regarding nutritious diets, workout plans, and active living were also examples of ways in which digital media supported health.

Relatedness

Digital media shaped many aspects of the relatedness domain and was a central component of the social lives of these youth. The one exception is the dearth of discussion about relationships with beloved pets – a very important consideration for the youth in the UNICEF Canada workshops (thus the report's name – *My cat makes me happy*). Participants used digital media to communicate with family and friends, to maintain "real life" friendships, to organize and plan activities, and to keep in touch over long distances. When balanced with offline communication, digital media use was perceived as a benefit to social well-being and created positive relations with peers:

> It's not that I wouldn't be happy without the technology, I just wouldn't be happy not being with my friends as much, not being in contact with them while we were away from each other ... I don't use technology often, but without it I feel like I would be a lot less involved with my friends than what I am because everything, if I make plans with my friends it's through text or it's on Facebook, it's not in person. (Iris, 18)

Being connected to one's online peer group helped to create a sense of belonging for some and was seen as a way to learn about new viewpoints and ideas. Rather than feeling addicted to digital media, some conceptualized it as being immersed with your peer network online: "It doesn't feel like, oh, I need to be on it; it's just like, I'm a part of it" (Gabriel, 20). This immersion sometimes came with a social responsibility to stay connected. Resisting digital media could result in missing out on elements of social life and planning with others.

Some participants felt that the attention they received on social media, such as "likes," boosted their self-esteem, but for others their relationship with social media was more difficult to manage. Participants described the drama and bullying they witnessed online and the pressure to feel accepted on, and gain popularity through, social media:

> If people stop talking to you, they won't like your pictures, it doesn't matter how good the picture is, they won't like your pictures because they don't want to talk to you ... It is a hard world, yeah, that's why I deactivated my Facebook. (Erika, 19)

Conscientious objectors who restricted their social media use still found alternative ways to communicate digitally (e.g., texting or emailing friends). Most participants acknowledged that if they did not engage in some form of digital media they would be excluded from planning and organizing events with friends: "It's like people don't know how to see you, I guess, other than online. Like, if they don't see you online, it seems like people forget about you a lot faster" (Carter, 18).

Interestingly some peer groups seemed to require a consistent public social media display of affection to signify that ongoing friendships were strong: "Looking back on it 20 years ago, people could still be friends without messaging each other every 20 minutes. And nowadays, to be friends with somebody, you need to Snapchat them at least once a day" (Jace, 17). Many participants noted there has been a decline in traditional forms of communication among youth, such as face-to-face and phone conversations, and described how mobile devices are now used as a shield against social anxiety.

Equity

While a number of participants could imagine that some other youth did not have cell phones or technology, most did not know such people directly. In imagining what it would be like, they spoke most often of people missing out on events or it being more difficult to communicate. Opting out of certain online spaces or the specific types of technology others use (e.g., iMessage, certain apps) means you can get left out of interactions and it can be harder to communicate with friends: "I think it leaves them out of what we now call, you know, the loop, you know what I mean, the social environments that we build online that everyone participates in" (Christopher, 17).

Youth also spoke about equity in online spaces in reference to encountering different people and ideas online. They recognized that you can control your image online so that you cannot necessarily tell who has more or less money, for example: "Like, as long as you have nice clothes on and your pictures are nice on Facebook, you're all equal there" (Erika, 19). This is considered good in that it can help you break free from negative experiences associated with being in a marginalized group (when you're online), but also problematic because youth acknowledge that it's not helpful to hide your reality in order to be accepted.

Young people also spoke about how some spaces online opened up the possibility of increasing visibility or acceptance of minorities and how social media has allowed them to connect with different forms of activism and social movements. Not all of these experiences have been positive (for example, the tendency for people with similar view points to stick together and to exclude others) but they do open up the possibility for learning and connecting with others who are fighting for equal rights: "without the Internet, I would've never known about the Uganda anti-homosexual bill, that's like right now, sitting on the president's desk. So, like those things,

I think that [the Internet] brings a lot of awareness" (Willow, 18). However, youth also mentioned that it is more common to encounter hate speech, racism, and sexism online:

> I think that they're definitely enhanced online because you're not face to face. So you don't think – you have something in between. You're not scared of expressing yourself fully, so you can say whatever you please. But then if you're in person you might just make a comment like under your breath or whatever ... It's definitely a lot more common to see online. (Madelyn, 16)

With respect to equity and young people's well-being, it is important to consider how it plays out in both physical and online spaces. It is important to attend to young people's experiences of (in)equity when it comes to access to technology and services needed to stay connected with others. It is also critical to understand the ways (in)equity can be experienced in online spaces and how that can shape everyday life.

Education and employment

Education is a common point of discussion in the interviews with youth. The 66 young people we refer to here reported limited use of technology within high school settings. Cell phones were often not allowed in class though there were a variety of different school or classroom rules to that effect, and differing levels of enforcement of rules by teachers. Most youth admitted to using cell phones in class, regardless of the rules. For most, using digital media in this context was distracting and not conducive to learning: "when I didn't have my phone I took more notes, understood more concepts, paid attention more" (David, 19).

Use of technology by high school teachers was limited to traditional uses (e.g., projectors, computers, PowerPoint presentations), and when youth did have classes about technology (e.g., media literacy) it was most often focussed on using computers and productivity software. Class content or police visits to educate youth about online safety and cyberbullying were also mentioned. Only a few students discussed taking media studies classes where there was more critical discussion and use of digital media (e.g., students writing a class blog). One youth shared that in media studies the students were asked to "analyze and critique media a lot and [they] also looked at how media does affect, like, individuals and society as a whole" (Sadie, 18). Youth generally felt there was not nearly enough education and quality curriculum or teaching relating to technology and about social media.

University and college students provided more examples of how technology was used in their education and classrooms. These included tools such as Moodle (an online learning platform, https://moodle.com/) or class websites, which relayed class content and were a means to submit assignments and communicate with instructors. Professors in these classes were perceived to be more accepting of the use of technology in class and, in some cases, tried to incorporate it into the class itself (e.g., encouraging tweeting of lecture material). In a few cases, youth had experience taking online courses or distance education programmes (both at high school and post-secondary levels). This wasn't always considered ideal by the youth:

> I don't like it [the e-learning course] that much. Being that it's being held in a different school board, I don't have direct access to the teacher. So any questions I have, have to be e-mailed, which takes time ... It's also a very different learning experience. (Easton, 17)

Despite this, online courses did provide educational opportunities youth would not have otherwise had.

More generally in terms of education, the majority of youth conversation was about how technology was beneficial for learning. It provides easier access to information through e-books and websites and can support independent learning: "I've learned, like, so many, like, historical things or something just, just by streaming on, just scrolling on the Internet or something ... when I'm on the Internet, I can learn to the best of my ability" (Stella, 19). While this access to information is mostly considered beneficial, youth also recognized that it can be an overwhelming amount of information:

> It can go both ways, you can find anything and everything on Google, you know what I mean, and that can be a great way to learn, but when you're on Google there's just so many other things and it can be easy to lose your train of thought. (Christopher, 17)

Some youth also believed that this ready access to information could hamper independent thinking and creativity since it is easy to find other people's ideas and cut/paste into your own work.

When it comes to employment or youth's future prospects, digital media use was again seen to have an important influence. For many young people, finding and applying for jobs happens online: "These days you don't just pop in and say I am looking for this job, a big career you are looking for, you send off an email, or call, or even videos, depending on what career you are going into" (Iris, 18). Additionally, youth spoke frequently about ensuring their digital presence was appropriate or professional (either now or in the near future). This entailed things such as building a professional image online, setting up an email account that doesn't use a nickname, and making sure that when their name is searched online it displays information they would want a future employer or university to see.

The topic of youth and precarious employment in the digital age is a critical aspect of youth studies as debates ensue around the extent to which automation and other aspects of Artificial Intelligence (AI) continue to replace young workers in global economies (Scully, 2016; Standing 2011, 2014). Our study provides limited input on this crucial question of youth well-being with future research needed on numbers, trends, and experiences of young people who are caught in precarious labour traps. In terms of both education and employment, attention must to be paid to the ways in which digital media and technology shape youth well-being through accessible, equitable, quality, and humanizing education and employment for young people.

Youth engagement

The concept of youth engagement vis-à-vis digital media was used by these youth participants in four interesting ways: (a) engaging in leisure and entertainment, (b) negotiating and cultivating identity, (c) engaging in cultural productions and process such as art and music, and (d) engaging in politics and political action. In the realm of leisure, digital media was, for many youth interviewed, an important means by which to relax and take time for themselves: "My phone ... that's how I relax, I just look on Facebook, look on Vine ... Pinterest" (Phyllis, 19). It also provides something to do when not with friends or at work or school: "it keeps me busy alone" (Easton, 17). Youth are able to use technology to explore their interests and passions and keep up to date on things important to them, including news, current events, activities of advocacy groups, gaming, hobbies, or lifestyle topics (e.g., active and healthy living, pregnancy, motherhood, etc.). Digital media provides youth opportunities to learn, share with others, and obtain advice on these areas of interest. One youth shared about the variety of interests that they can explore:

> My Tumblr's definitely my hobbies and my interests, I have two blogs of Tumblr that I connect, one's like 'fandom' stuff, so like TV shows I love, movies I love, actors I love, things are ridiculous and funny and make no sense - my second blog is books, writing, music, art that I enjoy. (Lucy, 16)

Another suggested that you could learn more about people you know in your everyday life when they share about their interests and hobbies online.

Youth also spoke about their use of digital media to actively learn about themselves and cultivate their identities. They spoke about choosing social media profile pictures to reflect their moods and feelings, and also how online identities and profiles help you define yourself:

> So if I post something, it's something that I wanted to post. If I want to share something, it's something I wanted to share that I find is important to me. Or whatever I'm following on Pinterest or Vine, stuff like that, it's what I find is important. (Phyllis, 19)

Youth felt that they grew along with the technology and learned more about themselves through using it: "I feel like as technology grew I did too, it helped me mature and learn new stuff ... I don't know, I just learned a lot from it, and it helped me be who I am, and like, grow" (Nevaeh, 20). Associated with cultivating identity is the idea that online spaces provide opportunities for youth to have a voice that they might not otherwise have in their everyday life:

> It does give us a voice, I will give you that. And I think that is why most people use it because a lot of kids ... what my perspective of on what they think is that their voice is not heard enough because they are teenagers who gives a crap, right? So, they go on a computer and all of a sudden they have a voice and they get all these likes and they to keep posting. It really gives them confidence. (Mason, 17)

It can be easier to express oneself online, especially for those who are more shy or introverted, and some youth shared that it's easier to be yourself and not feel judged.

Stemming from the ways in which online spaces allow for youth voice and exploration of interests is the fact that youth can more easily share their own cultural productions and creations online. Youth participants told us about using digital media to promote themselves as artists (e.g., writing samples, performance videos, music samples) and how they used digital media to create professional images and portfolios. Not only could they share their creativity online, but they also felt digital media use increased their creativity: "I really think that the most massive part of my life that technology influences is that I consider myself to be rather creative and I feel like I can use the Internet to think more and be more creative" (Chase, 17).

Finally, there were some forms of political engagement and activity made possible by digital media. For some youth, the Internet made it possible to follow the news and actions of organizations of interest: "I get emails all the time from Amnesty International and from PETA, so it's nice to stay in the loop with what's going on there" (Willow, 18). Others took a more active role in spreading information online: "if there are things that I find like close to my heart – then I'll definitely post something, like 'Hey, sign this petition about the ... pipeline'" (Violet, 17). In summary, digital media was reported to be playing important roles in providing youth engagement opportunities in the personal, social, and political realms. The question remains as to whether and how this engagement could be expanded with decidedly youth-based tools, platforms, and designs with the express intention for furthering youth engagement.

Affordable living conditions

Access to affordable online and mobile services remains a challenge for some of these young people. Participants did consider the cost of data plans and devices to be very expensive. Parents and guardians often paid these costs, while some participants worked and paid themselves:

> It's hard to keep up with all the new technology on the price that these things come out with and it's crazy ... I don't manage it that well, right now I'm still a dependant at home, you know what I mean, so my parents are paying for everything. (Christopher, 17)

The social pressure to keep up with technology made it difficult for those who could not afford expensive models of phones and data plans. Participants who did not have data plans relied on Wi-Fi for digital communication and information. When participants did not have access to cellular data, they would often use more "traditional" media such as landlines or computers for communication. If data or Internet was not available at home, public Wi-Fi seemed accessible in urban areas. The gap between those who did and did not have access did not seem insurmountable given the number of access points to Wi-Fi and types of devices that message using free app software; however it still created social barriers and divisions in communications with peers. Having affordable access to mobile and digital media should be viewed as an important component of affordable living domain of well-being for youth to actively participate in their social networks. Chapter 6 provides a more detailed explication of how young people could gain access to necessary material resources and economic opportunity as a part of the digital capital that is the foundation of this domain.

Space and environment

One aspect of this domain involves youth having positive environments and living conditions. Youth well-being in the digital age must also take digital ecologies into account. The most predominant motif from youth interviews in relation to this domain had to do with safety in online spaces. Youth spoke about a variety of problems including catfishing schemes, cyberbullying, online harassment, and the Internet as a space that "feeds" terrorist groups. Most frequently, their comments about safety had to do with how much information they provided about themselves online: not posting too many pictures because it wasn't safe, not posting about vacations since people would know you are not home and could lead to home invasions, and limiting what information they share online because of how easy it is to be "stalked." Many youth were cognizant of such problems from media reports (e.g., Amanda Todd[2]) and police visits to their schools to discuss cyber safety. Some young people also discussed digital media safety in regard to government or corporate surveillance, most often pointing out that it's "creepy" when social media sites such as Facebook tailor advertisements based on Internet searches previously conducted, or how others know where you are if you have GPS enabled on your electronic devices.

A healthy natural environment is also key to this domain of well-being and was often equated with experiencing the outdoors and nature. Youth suggested that using technology often takes away from going outside:

> I think just losing touch with what the world really is. 'Cause a lot of us now, these days, the only way you see the world is through a screen and through social media ... And you don't, you forget what it's like to look out your window and know that that's the world too, you know what I mean? (David, 19)

Another participant expressed how she felt before being immersed in a digital environment:

> When I was a kid I had so much imagination. Like, go outside and run through the wood and like go adventuring, go look to see what I could find in the woods and now it's like, sit at home, play my Xbox, be lazy, play, play games on Facebook or something. (Rose, 18)

Thus, when considering well-being in terms of space and environment, understanding how online ecologies (both the tools and life forms, see Chapter 6) are experienced as (un)safe is key. In addition, we acknowledge the ways in which youth are experiencing and attempting to navigate the balance of online/digital and offline/nature in young lives. Finding balance was a crucial part of youth conversations regardless of the motif they discussed. It is therefore considered here as a key aspect to a new domain of youth well-being, as discussed in the following section.

A new domain: digital lives

While much of what youth told us about their digital media use can be associated with the seven domains of well-being discussed earlier, there were four additional themes identified that are pertinent to youth well-being yet do not fit easily into the existing domains: consumerism, shifting sense of time, dependency on technology, and finding digital balance. We outline these themes, in turn, and suggest the need to expand our understanding and measurement of them within what is a useful new domain: digital lives.

Consumerism

A common point of discussion across many youth interviews was the connection between digital media, marketing, and consumerism as a daily feature of digital capitalism. Chapter 6 expands further on these ideas. The 66 youth interviews analyzed here pointed out that advertising is a pervasive impetus in online spaces, including on Facebook and YouTube. Youth participants reported that they felt targeted when their Internet searches were used to populate the advertising spaces on websites and in apps. Some youth also recognized that not only were advertisers selling to the youth market through online advertisements, but the technology itself was being marketed to youth, creating a ubiquitous culture of technological consumption:

> Most of their apps and social media apps are geared towards our age group ... I feel like we're just the most vulnerable crowd for them to zone in on and for them to get as much as they possibly can, out of us, for their benefit. (Naomi, 20)

Youth also recognized the built-in obsolescence of technology and that "even if your thing is the fastest, greatest thing on the market, in two years, there's going to be something that's even faster and your thing is going to be slower than that" (Jace, 17).

Some youth felt that the large technology companies, who are focussed on making money, do not care about youth, or society generally. Some expressed that the companies are "making money from stuff that means nothing to anyone but is still captivating and addicting" (Naomi, 20) and that "a lot of the point of technology is to make money off of wasting people's time" (Easton, 17). It is important to understand how this pervasiveness of capitalistic aspects of consumer culture and marketing/targeting of youth impacts their well-being when it is so prevalent in their everyday actions and interactions. Measures of levels of youth consumerism, both offline and online; amount of marketing and advertising encountered; trends in personal data breaches; and the forms of tools and designs to encourage youth to become addicted to apps could be monitored.

Shifting sense of time

Another influence of digital media on youth well-being is the experience of a shifting sense of time. Some youth spoke to this theme as feeling that time seems to melt as it goes past quickly and disappears when they are engaged in online spaces and activities. There was also an increasing sense of "wasting time" when using digital media, despite the fact that these activities can be integral to interpersonal connections and developing youth identities. Technology can be ubiquitous in young lives and results in time and experience seeming to pass very quickly. Ideas, words, pictures, and information all spread rapidly through digital media and make it difficult to keep up. This leads to feelings of discomfort or unease: "I feel like we're so used to like, things being fast and easily available and ... just like really rushed. Like, it makes everything rushed" (Elijah, 19).

Despite events happening quickly, there is also a sense of problematic permanence in online spaces. Youth are keenly aware that words and pictures they post online can be saved, screenshotted, shared, and otherwise spread even if they are later deleted:

> When you put something out there, when you say something, it can't easily be taken back because someone's seen it. There's always that saying, 'If it's on the Internet, it's there forever.' And I believe that that's really true. Even if you delete it, someone's seen it and the messages still got across. So getting yourself into trouble is much easier in that regard. (Easton, 17)

The fact that whatever you share electronically could continue to exist and affect you or your image/reputation was a concern for some: "If you delete something, it's not actually deleted or whatever. So that's pretty scary" (Madelyn, 16).

Dependency on technology

Youth expressed concerns about becoming dependent on technology. In one respect it has become integrated into their means of communication:

> All somebody's got to do is text somebody, or call somebody. Like it's good, because I mean it makes life easier instead of having to go walk to go see ... somebody, but at the same time it's like we're, we're too reliant. (Rose, 18)

In another sense, as discussed earlier, youth shared that they were becoming accustomed to always being connected (to information, to people, etc.) and have a difficult time disconnecting. For some it was a matter of feeling "like you've kind of got to check up on that stuff [online]" (Carter, 18) or feeling stressed if they don't have their phone. Some young people expressed a sense of loss of control when it came to their constant use of digital media: "It's like so addictive ... you can't control it. Once you start, you can't control it" (Erika, 19). This sense of dependency, reliance, and loss of control (or addiction) is shaping young lives and their sense of well-being. Further understandings and measures of forms of digital life such as ubiquity, dependence, and speed of digital ecologies are required.

Finding digital balance

Finally, the overarching motif that repeated throughout youth interviews was the challenge of seeking, finding, and maintaining balance or harmony in their lives with respect to digital media. This theme arises repeatedly through this book and is embedded in all of the proposed new themes described earlier. There is a pervasive sense that technological change has happened so quickly that young people have not yet figured out how best to integrate digital media use into their everyday lives – particularly in healthy ways. One youth said, "I think it's happening so quickly that no one knows what [is] the right thing to do" (Naomi, 20), and another pointed out that:

> No one taught anyone how to use the Internet. There wasn't, like even when this was first made, there wasn't like a course, you know? There wasn't, it was just thrown at you and say, 'Here, here ya go. Make something'. (Elijah, 19)

Young people tried to distinguish between their everyday lives/selves and their online lives/selves, but acknowledged "it's so mixed together that you can't break it apart" (Elena, 17). Easton (17) shared:

> I think just technology is – I don't want to say everything because it's really not. It just takes up a lot of people's lives, my life too. Like, everything I do is technology. Like, I'm never without my phone; I'm never – yeah. There's usually some sort of technology following me everywhere, so I guess knowing that it is kind of a way of life, technology. It's becoming a way of life. Not a bad thing, but ... I don't really have an opinion on whether it's ruining everything or not because I'm kind of living with it either way.

Many expressed having a love/hate relationship with digital media:

> I guess I like it because I get to see what's out there kind of and keep in contact with people. But I also hate it because I don't get that interaction anymore and it kind of is time consuming. (Phyllis, 19)

Trying to come to terms with how much time they spend on digital media is a challenge many youth mentioned. They appreciate the conveniences offered by the technology and understand why they use it so frequently but also felt that there is an issue of overuse in their lives and in broader society:

> So I think some people can get obsessive with it but really I think people just need to be taught like ... you can go on Facebook, you can go on

Twitter, you can go on Skype and Tumblr and Instagram, you know, just don't spend your life there. (Elijah, 19)

For other youth, they could tell that too much digital media use was not good for their well-being: "You obviously want to have it but, like, sometimes what you want is not what you need" (Esther, 18).

Summary and directions

After weaving together the stories and experiences of the Canadian youth participants who spoke about the interrelations between digital media and well-being, we are in a position to suggest a model of youth well-being for the digital age. In summary, we have shown that digital media use in young lives is integral to their sense of well-being, influencing all of the established domains of youth well-being as proposed in the UNICEF Canada (n.d.) CY-Index. However, youth well-being must include additional domains and themes with respect to digital media, which must be more explicitly addressed in conceptualization and measurement.

In this chapter, we asked: How does digital media use figure into each of the domains of well-being? What considerations are essential for understanding well-being in the context of young lives saturated by digital media use? Do the domains fully encompass aspects of youth well-being in the digital age? Is there anything missing? Youth well-being models to date have not reckoned with the pervasive encroachment of digital media into young lives. Overconsumption and unhealthy use of digital media negatively impact well-being for youth engagement, social relatedness, and education. Yet some digital presence/literacy is needed to guide youth activities, contribution in peer spaces, maintaining relationships, etc. For example, as quoted earlier, excessive use of digital media was associated with "losing touch with what the world really is" (David, 19), yet digital media use is also associated with fostering relationships and knowledge about the world around us. A youth well-being model must consider healthy use and integration of digital media in young lives and manage the contradictions experienced by youth. We propose the following as possible additions to a model of youth well-being based on what youth told us of their lives:

As reported in this chapter and illustrated in Table 3.1, youth well-being relies on a balanced approach to digital media and such considerations must be included in a relevant index. Indigenous models of well-being have much value to add to this notion of balance that is sought by people and communities, based in the balance of medicine wheels to include physical, spiritual, emotional, and intellectual aspects (Ontario's Mental Health & Addictions Leadership Advisory Council, 2015; Toulouse, 2016). In these models, the notion of hope, purpose, belonging, and meaning ground well-being and are

Table 3.1 Considerations for a Model of Youth Well-Being in the Digital Age

Domain	Existing Digital Media Indicator[1]	Additional Digital Media Themes	Additional Digital Media Indicators
Health	n/a	Digital media access and practices associated with mental and physical health Digital media access and practices associated with spiritual and reproductive health	Consumption and exposure to commercial media content Addiction/dependency Online hatred, violence, violent content, and bullying Accessing healthcare/information online
Relatedness	Cyberbullying Access to Internet and social media	Pervasiveness of digital media through all domains of relatedness Balance between online/offline self	Online relationship drama Mediated social interactions Relationship skills – offline and online Balancing social obligations Healthy online relationships Cultivating meaningful online identity
Equity	Diverse representation in media and decision-making roles	Marginalization due to lack of digital access/skills Discrimination online Access to quality technology and digital media literacy education	Access to digital media (measures of the "digital divide") Skills required for effective use of digital media (measures of the "second digital divide" – skills and competence) Equity of access and opportunity Hatred/discrimination online

(Continued)

Domain	Existing Digital Media Indicator[1]	Additional Digital Media Themes	Additional Digital Media Indicators
Education and employment	n/a	Distraction from education and employment pursuits Digital competencies for education and employment	Distraction and ability to focus on tasks Digital skills for education (research, assessment, synthesis, etc.) Digital skills for employment (job seeking, self-promotion, etc.)
Youth engagement	Accessible online spaces Inclusive media	Most activities, decisions, spaces are mediated online (to some degree)	Online platforms to engage youth Youth contributions to online platforms Online opportunities to explore identity Youth digital media production/ creation Online hatred/ bullying resulting in silencing
Affordable living conditions	Access to Wi-Fi	Digital access as a material condition of affordable living	Monthly plans and devices affordable Accessible Wi-Fi available Digital technologies available Digital Capital measures (social, symbolic, cultural, monetary – see Chapter 6)
Space and environment	n/a	Depth and complexity of digital ecologies Relationship between digital and physical spaces	Amount and character of use of (1) physical spaces (nature) and (2) digital ecologies (tools and life forms) Time management skills Online safety Surveillance Privacy online

Domain	Existing Digital Media Indicator[1]	Additional Digital Media Themes	Additional Digital Media Indicators
Digital lives (new domain)	n/a	Prevalence of advertising and consumerist discourse Shifting sense of time Dependency on technology Balancing online/offline and physical/digital lives	Time management skills Healthy relationship skills Healthy social interactions/communication (online and off) Digital literacies Safe online practices Exposure to advertising and marketing Social pressures to be available (in communication) Ubiquity of digital media Speed of content spread and communication

[1] Adapted and developed based on the model presented by UNICEF Canada and Students Commission of Canada, 2017.

thus suggested here as a strong basis of a holistic model of youth well-being in the digital age (see Figure 3.1).

Hope is developed in the spiritual domain. Hope is sought in values, beliefs, and identity, and in being aware of self and social situations. The emotional domain is related in that it is behaviour expressed through relationships from which *belonging* comes. The intellectual domain is rational, thoughtful, intuitive, and full of understanding and decision-making that lead to *meaning*. The physical domain develops holistic *purpose* through behaviours of being and doing in the wholeness of the planet, community, family, home, and self. When the young person finds hope, purpose, belonging, and meaning, they are said to have well-being. Thus, each state must be negotiated and renegotiated across the domains of well-being. When young people access, encounter, know, and negotiate digital ecologies, they encounter both risks and opportunities for their well-being. Our model contributes by asking how digital media influences youth well-being across each of the domains and themes. It is guided by the questions: How do youth negotiate the risks and opportunities provided by digital technologies and ecologies? How does each encounter provide or negate the hope, purpose, belonging, and meaning required for flourishing well-being? To what extent do young people know and negotiate their access, skills, risks, and opportunities on a

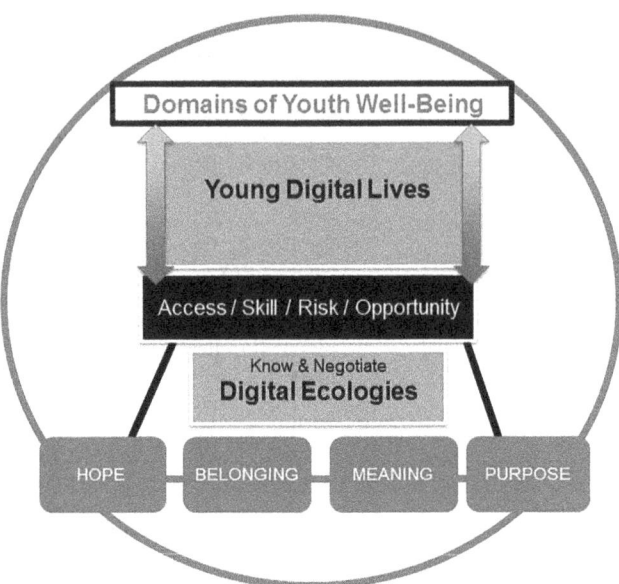

Figure 3.1 Domains of Youth Well-Being.

daily, monthly, and annual basis? How do they find balance in their digital ecologies? A more holistic and robust model of youth well-being in the digital age requires that we attend to such questions and work together to find better concepts and measures. "Would I be less anxious without my phone? Maybe, maybe I would feel more at peace or maybe I would feel more separated, I actually can't say because it feels like it is physically a part of me, you know" (Gavin, 19).

Notes

1 All participants were given a pseudonym.
2 Amanda Todd was a 15-year-old who had experienced bullying (online and offline) and ultimately committed suicide, http://www.amandatoddlegacy.org/

References

Association of Ontario Health Centres. (n.d.) The Canadian Index of Wellbeing: A made-in-Canada tool to measure what matters. Retrieved from http://communityhealthandwellbeing.org/canadian-index-wellbeing
Azzopardi, P., Kennedy, E., & Patton, G. C. (2017). *Data and indicators to measure adolescent health, social development and well-being.* (Innocenti Research Brief 2017-04). Florence, Italy: United Nations International Children's Emergency Fund. Retrieved from https://www.unicef-irc.org/publications/pdf/IRB_2017_04_Adol02.pdf

Canadian Index of Wellbeing. (2016). *How are Canadians really doing? The 2016 CIW national report.* Waterloo, ON: Canadian Index of Wellbeing and University of Waterloo.

Michalos, A. C., Smale, B., Labonté, R., Muharjarine, N., Scott, K., Moore, K., ... Hyman, I. (2011). *The Canadian Index of Wellbeing.* (Technical Report 1.0). Waterloo, ON: Canadian Index of Wellbeing and University of Waterloo. Retrieved from https://uwaterloo.ca/canadian-index-wellbeing/sites/ca.canadian-index-wellbeing/files/uploads/files/Canadian_Index_of_Wellbeing-TechnicalPaper-FINAL.pdf

Ontario's Mental Health & Addictions Leadership Advisory Council. (2015). *Better mental health means better health.* Retrieved from https://static1.squarespace.com/static/5845afbfbebafb2a2ebd4321/t/5848532c893fc0d86fc71af7/1481134893744/mental_health_adv_council%281%29.pdf

Scully, B. (2016). Precarity north and south: A southern critique of Guy Standing. *Global Labour Journal, 7*(2), 160–173.

Standing, G. (2011). *The precariat: The new dangerous class.* London, UK: Bloomsbury Academic.

Standing, G. (2014). *A precariat charter: From denizens to citizens.* London, UK: Bloomsbury Academic.

Toulouse, P. R. (2016). *What matters in Indigenous education: Implementing a vision committed to holism, diversity and engagement.* Toronto, ON: Measuring What Matters, People for Education. Retrieved from http://peopleforeducation.ca/wp-content/uploads/2017/07/MWM-What-Matters-in-Indigenous-Education.pdf

UNICEF Canada. (n.d.) *Measuring well-being.* Retrieved from https://www.unicef.ca/one-youth/child-and-youth-well-being-index/

UNICEF Canada. (2017). *UNICEF report card 14: Canadian companion, Oh Canada! our kids deserve better.* Toronto, ON: UNICEF Canada. Retrieved from https://www.unicef.ca/sites/default/files/2017-06/RC14%20Canadian%20Companion_0.pdf

UNICEF Canada & Students Commission of Canada. (2017). *My cat makes me happy: What children and youth say about measuring their well-being.* Toronto, ON: UNICEF Canada. Retrieved from http://www.unicef.ca/sites/default/files/2017-08/UNICEF_One%20Youth%20Report.pdf

Young Lives, U. K. (2017, Dec). New measures of adolescent development: Workshop report. Paper presented at the *CIFAR Research Workshop,* Oxford, UK.

Chapter 4

The way we live now

Privacy, surveillance, and control
of youth in the digital age

*Heather Barnick, Valerie M. Campbell,
and Kate C. Tilleczek*

Introduction: youth, digital technology, and culture

> you have to have some form of technology, social media, or you're going to be this ghost that no one sees ever[1] (Elena[2], 17).

> It's like we're not actually free from the technology ... I find that it's strange, like you're trying to fit in, almost, with society. But we are a society who's doing this to ourselves (Madelyn, 16).

> I feel like technology, or media and all that stuff, controls who we are and the values that we hold ... sometimes technology or media can influence some of the things that we do ... in the future I would like it to be the opposite. (Sadie, 18)

This chapter provides an overview of how young people think about and respond to problems of privacy, surveillance, and control when they go online or use digital technologies. In so doing, we recount the structural and social relations through which agency and control are produced in young digital lives. The perspectives provided are those of 145 participants from Canada, Scotland, and Australia who were in their late teens or early twenties when they were interviewed (between 2013 and 2015). Their views about privacy, surveillance, and control are situated within the macro-level processes through which state and corporate interests have consolidated and colonized cyberspace thereby posing serious risks to the ability of ordinary citizens to exercise democratic restraints on state and corporate power (Dayen, 2017).

We focus on participants' concerns about and efforts to control their privacy and attempt to explain their tacit acceptance of surveillance as well as the reasons they do not exercise the means of control available to them, such as reading terms of service (ToS) or refusing to use certain digital devices and social media applications altogether. Elena, Madelyn, and Sadie's statements, quoted earlier, provide a good starting place to understand why young people so often express feelings of resignation towards surveillance and loss of privacy. All three points of view draw attention to important shifts in the ways in which young people are relating to digital technologies. More than external devices or tools, youth participants often spoke about digital technologies as vital components of a whole way of life. For Elena, deciding to use technology and social media was not a legitimate choice at all; one's very existence could be at stake because without technology, you could become a "ghost that no one sees." Some, like Sadie, were also aware that modern technologies are forces of control, not only because they are designed by and for governments or Big Tech[3] but also because they are infused with and infusing how we think, how we feel, what we do, the values we hold, and even who we are.

Participants were divided over whether this infusion is good or bad, but there was a general consensus among them that digital technologies are not just part of contemporary culture, they *are* that culture – firmly embedded in the meanings, practices, and material conditions that shape everyday life. While reflecting on this, David (19) compared digital technologies to religion and, in his view, trying to change people's attitudes and behaviours towards them would be futile as promoting atheism to a "diehard Christian": "it's hard to change a lot of people's philosophies and views … just like if you tell a diehard Christian that God's not real." Perhaps because digital technologies have become constitutive of modern-day culture, it is easy to forget that they are not just for social networking, communicating, and accessing information. They are also designed for data tracking and extraction, twin processes that are constantly infiltrating "free" time and "private" space while also manipulating thoughts, feelings, and actions in ways that are beneficial to governments and corporations, but often detrimental to civil liberties (Dayen, 2017).

Another fact that is often overlooked is that modern devices are composites of analogue and digital technologies and this means control over personal information is not only about what is posted or shared on the Internet. Most devices like smartphones, tablets, and iPads no longer need a Wi-Fi connection to invade our privacy (Newman, 2017). Even when devices are offline, they are "always on," gathering, circulating, and learning from the data we post online and even the data we do not (Laboisse, 2017). Privacy debates often neglect the analogue sensors attached to our digital devices that register information about air pressure, humidity, temperature, speed, light, motion, and heart rate (Laboisse, 2017). As researchers at Princeton have demonstrated, "seemingly innocuous sensors can be exploited using machine-learning techniques to infer sensitive details about our lives" (Mittal as cited in Shekhtman, 2017, para. 4). Many participants were aware of the ways their personal information was vulnerable through their activities online, but the fact that their data could be intercepted when they were offline was an issue few of them discussed.

The inability to track what happens to one's personal online data (where it goes, who gets it) and to know when personal data are being tracked are two of the most insidious forms of technological control today. As Weber and Wong (2017) explained, the core principles guiding regulatory limitations on data collection and privacy protection do not translate well in the context of the Internet of Things (IoT), a term that when broadly defined refers to the connection of devices (i.e., speakers) or ordinary physical objects (i.e., kitchen appliances) to the Internet (Meola, 2018, "What is the Internet of Things?" para. 1). The principle of "data minimization," for instance, restricts data collection to information that is directly relevant to the service offered; however, in IoT applications one piece of data might have multiple functions, and "the real promise lies in combinations and permutations of data that multiply insight in sometimes surprising ways" (Weber & Wong, 2017, "Privacy debates will center," para. 3). Similar issues arise with "notice and choice" regulations that guide transparency laws governing privacy policies (PPs) and ToS contracts (Weber & Wong, 2017, "Privacy debates will center," para. 5). These laws are designed to empower users to make decisions about access to their personal data. However, users who are enticed by "click to agree" contracts rarely read PPs, End-User Licensing Agreements, or ToS, which are often long, jargon-filled, and full of small print (Ayres & Schwartz, 2014; Berreby, 2017; Obar & Oeldorf-Hirsch, 2016).

Another issue at the intersections of consumerism and surveillance is geo-targeted mobile marketing, accomplished when businesses and advertisers access the GPS chips in our phones allowing them to use "pings"[4] to track our daily foot traffic and create a demographic profile based on where we go, with whom, and how long we stay (Newman, 2017, para. 2). Some geo-location technologies can function even without a phone's GPS. Beacons placed outside stores can use ultrasonic signals to communicate with our phones, even when they are in our pockets and even when Wi-Fi and GPS functions have been disabled (Newman, 2017, para. 2).

The surveillance capabilities of our modern technologies compound data flow issues in privacy debates by creating legal loopholes that can be exploited by government agencies or private companies. These loopholes often open up in scenarios when technologies communicate with each other about human activity, but in their own techno-language (numbers and codes). Mic-tapping applications, like those embedded in *Amazon Echo*, for example, perfectly comply with PPs because they do not listen in on human conversations or record human speech, only ultrasonic signals from commercial audio sources (Sulleyman, 2017). Our participants were savvy about the fact that they were constantly being monitored and tracked through their digital technologies – "they [Government] can track all your stuff, everything that you do" (Charles, 19) – but because topics like beacons and smartphones sensors did not emerge in our conversations with them, it is unclear how many were aware of the analogue technologies through which this could occur. Similarly, while many participants were quite knowledgeable about website or routing options that allowed them to tighten their privacy settings – "there's [sic] undercurrents of people who are using Tor, which is an Internet browser that makes you completely anonymous" (Carolyn, 17) – they were generally less astute about actual laws or regulations related to their right to privacy. The fact that many contemporary privacy regulations continue to lag far behind the pace of technological change is an important mechanism of control in the 21st century. This lag creates ambiguities that make it easier for the State and Big Tech to define the law according to their interests while ordinary citizens find it more difficult to appeal to their rights as a defence in court proceedings (cf. ACLU Staff, 2017; Ayres & Schwartz, 2014; Chan, 2017).

Theorizing control in the digital age

Young people's attitudes and behaviours towards technology as a non-negotiable way of life pose many challenges for contemporary social scientists seeking to understand how the IoT becomes an ecology to negotiate as it links digital devices, capitalism, and social life to systems of ethics, power, and control (Coleman, 2009; Dyer-Witherford, 2015). A particularly useful insight from Karl Marx's work was that capitalism was not a fixed mode of production but one that would continue to develop alongside corresponding alterations to the logic of "productivity," which insists the value of products always exceeds the cost of production (Hall, 1977, p. 43). Elaborating on Marx, Gramsci developed the concept of "hegemony" to understand how mass consent to systems of domination occurs as the productive logic of capitalism insinuates itself so firmly in social relations, institutions, and ideologies that it becomes a totalizing culture (Hall, 1977, pp. 65–66). Gramsci's writings led several scholars to reread the relationship between Marx's base (material conditions and forces of production) and superstructure (political and legal institutions and ideologies) and to look for analyses that keep social being, ideological being, and material existence simultaneously in view (Althusser, 1971; Hall, 1977; Williams, 1977). This flattening of Marx's base and superstructure provides a starting place for thinking of power and control "cybernetically," that is, as a self-regulating system in which behaviour is continuously modified by feedback-loops of circulating information.

The application of Marxist theory to contemporary "techno-capitalism" (Dyer-Witheford, 1999, 2015) has been fruitful for thinking about how the Internet has structured a global system of class relations and exploitation, along with new mechanisms for generating surplus value (Burston, Dyer-Witherford, & Hearn, 2010; Fuchs, 2014; Grossberg, 2010). In this chapter we keep Marxism in view but draw on insights from Deleuze's short essay "Postscripts on the Societies of Control" (1992), in which he characterizes late capitalism as a "a generalized crisis in relation to all environments of enclosure" (p. 1). On this point Deleuze provides a useful addition to Marx with his observations about how societies in late capitalism have initiated a new set of processes for ensuring that individual, collective, and institutional bodies always behave productively. The "crisis" Deleuze outlines provides a helpful frame for thinking through our participants' descriptions of power and control in terms of time (pressure to keep up), space (always being watched), and place (feeling lost or left out without technology).

For the purposes of this chapter two particular insights from Deleuze are important.

First, "societies of control" mark the growth of a new corporate logic that operates without spatial or temporal limits. Building from the work of Michel Foucault, Deleuze observes how control over a population could be achieved by continuously "modulating" rather than "molding" human and non-human bodies. As explained by Naomi (20):

> Twenty-year-olds don't know who they are. They still want all that powerful energy that you get from being behind a screen. Posting whatever and getting likes on Instagram and getting the recognition. Feeling the power, the temporary power for like a day … Causes a lot of unnecessary thought patterns that then stick and stay and continue and linger and then that becomes a habit and then you're kind of stuck with it.

Within this next phase of modulatory capitalism, Deleuze argues power and control can no longer be understood only as the effects of the disciplinary practices which Foucault described as "biopower" (1990, 1995). Instead, what becomes more important are the systems and hierarchies that govern over informational access and action (Deleuze, 1992, p. 5).

In temporal terms, the persistent use of digital technologies ensures that there is less leisure time (Asano, 2017) and that all space/times are potentially productive; even when users watch television, ride the subway, or fall asleep, their mobile devices are able to gather valuable data about them (Nanos, 2018). At the same time, spatial shifts occur with the growth of global networks that constitute the World Wide Web and the more localized networks that link human bodies, the surrounding environment, and multiple devices into the IoT.

For Madelyn (16) this provoked anxiety and confusion:

> the mindset that we need the newest thing, we need all this technology in the world to be in our hands at anytime that we ever want. So if we realize then, well I think there might even be laws regarding how often or whatever … But then if people don't realize, then it'll just, I feel like on a biology level, we might start changing our body to fit the technology, which is kind of freaky.

Madelyn's concerns reflect how the combination of these two kinds of networks have undermined the categories that once marked physical boundaries and the corresponding legal jurisdictions that place limits on state and corporate infiltration into civilians' lives – foreign/domestic, suspect/citizen, the privacy of one's home/the openness of public space.[5]

The danger in the collapse of these categories is that it offers the state and corporations the opportunity for manoeuvring around due-process laws and constitutional rights that would normally place procedural and territorial limits on acquisition of personal information.[6] As Deleuze predicted, when these categories of enclosure are undermined, the law becomes increasingly "hesitant" while the judicial system as a whole finds itself in a "crisis" (Deleuze, 1992, p. 5).

The second insight from Deleuze is his collapsing of the mass/individual pair. This reorients the subjective experience of individuals as imminent, never bounded by a fixed and stable identity position, but always open and in a state of becoming. Another important temporal dimension of a society with no outside is that "one is never finished with anything" including one-self (Deleuze, 1992, p. 5). We observed this in the way several participants spoke about the need to constantly adjust their digital footprint to create the right impression. For some the maintenance of their social media profiles was a constant obsession. A notable example was Willow (18), who learned during a digital media course how to polish her online impression by joining professional networking sites "so all the bad stuff gets pushed back." For her, managing one's digital footprint is a never-ending process because "you can't get rid of anything that's online, but you can bury it deep enough." For others, creating the right impression was less about burying incriminating content and more about competing with others. Naomi (20), a dancer, spoke about the anxiety this caused her because of the need to compete with others in her profession: "I compare myself to them and that becomes discouraging, so in the end it's like an inner battle, but it stems from the competition that is out there."

Both the character and designs of social media applications and social networks, which measure users in terms of friends, likes, or followers, compel individuals to constantly update, alter, reinvent, and compare themselves with others. Deleuze understood this kind of "competition" as a corporate logic insinuating itself into non-corporate spaces. Competition is another way digital technologies deploy modulatory control as a motivational force that pits individuals against each other and divides them from within (Deleuze, 1992, p. 5).

How does modulatory control function?

Modulation is the process through which any thing, person, or system can be rendered as information and carried as code (Deleuze, 1992). The concept is useful for understanding control as an effect of information production processes working through the designs of digital technology to place boundaries on experience. One such process includes the gathering of user data through background programs (i.e., cookies) that track search histories and other online activity. These invisible programs exert control by customizing users' online experience so that it is tailored to suit the specific tastes, preferences, and values of each individual. Not many participants were thinking about these processes in terms of control, but a few were aware of how filtering could control their worldviews and perceptions by restricting their access to information: Adeline (18) saw Google's search engine as limiting because it prioritized what she searched most often or what was most popular – "it's constraining because it's kind of blinding you from other sources of information." Bailey (21) spoke about a "filter bubble" that was hard to escape: "You're in a bubble, just your information ... Because it's filtered for your individual needs and your search history."

Another important process to consider is the labour of users, who collectively could be thought of as a kind of proletariat, especially since the utopian promises of digital technologies as articulated by companies like Facebook and Google are addressed not to clients, or guests, but to users. What we must remember though is that these users do not receive a wage, yet produce *themselves* as data in their free-time, as they post, share, like, follow, watch, shop, and even when they sleep. As Big Tech companies capitalize on unpaid user labour, which they collect by offering so-called free services, their power to influence the global economy has grown into what has been dubbed "the new predatory capitalism" (Dayen, 2017). User-generated data are one of the most valuable commodities in this form of capitalism and some estimates place its worth at over $300 billion in the financial sector alone (Strategy & cited in Short & Todd, 2017, "Data as strategic asset," para. 2). According to an article posted in *Forbes* magazine, nearly all companies doing business online acknowledge that "big data" is as valuable a revenue source as their actual products and services (Baldwin, 2015, para. 5).

Sadie (18) was one of many participants who were aware that free-time spent online should not be confused with freedom, though she also pointed out it is sometimes easy to confuse the two because the Internet can feel like an escape from reality "For me, it's just ... I guess it's just something, fun to do. It's, free time. It's just, to get away from, I guess, reality." Yet, Sadie was less optimistic about how much freedom she had when it came to protecting her privacy:

> I guess we do have freedom but, not really because on my phone I have to change all the settings because every single app has access to your location. So it's tracking you like every step ... it can tell you where you are right now.

Some participants reflected on how the economic might of Big Tech afforded them power to impact other spheres. David (19), for one, was troubled by the capacity of Big Tech companies to influence the market as well as social values:

> big companies like Apple or Google, their stocks are big on the DOW Jones so ... they're pretty big influences on the economy ... it's like a brainwash almost ... this is the only way kinda thing ... It's kind of like saying this is the only virtuous path.

The next section elaborates on these processes of modulatory control by providing a historical context for participants' points of view on privacy and surveillance. What these narratives reveal is how participants' attitudes and responses to issues around privacy and surveillance are connected to the force of modulatory control, which they experienced as social pressure (keeping up with peers, keeping up with technology) and disorientation (feeling lost, left out, or confused without technology).

How digital technologies became a form of life

The first decade of the 21st century saw the introduction and growth of social media and mobile digital devices to an extent that was unprecedented, and perhaps unpredictable. Consumer reports from Canada, for example, show that cell phone ownership for personal use doubled or tripled in the period between 1997 and 2004 (Industry Canada, 2006). Even at these rates, Canada lagged behind most Organization for Economic Co-operation and Development (OECD) countries where the average number of wireless communications subscribers had reached 53 out of 100 subscribers four years earlier (Industry Canada, 2006).

As early as 1992, Multicast Backbone (MBone) protocols made video and audio casting over the Internet possible (Almeroth, 1999). This technology was initially limited to a small circle of computer scientists, engineers, or academics, but eventually gained mass popularity with introduction of Skype in 2003. The global sharing of user-generated content quickly followed with sites like YouTube (2005), Tumblr (2007), and Snapchat (2011). Meanwhile, media apps such as MSN Messenger (1999), Friendster (2002), and Myspace (2003) were becoming popular within specific demographics. This explosion of social activity on the Internet occurred alongside the introduction of mobile Personal Digital Assistants (PDAs) like the Blackberry and later the iPhone and iPad. These were some of the earliest "convergence devices" (Jenkins, 2006) capable of telecommunications, Internet connectivity, GPS mapping, and even biometric readings. But it was the introduction of Facebook in 2005 that heralded the breakthrough of social media into a much wider arena (Ahn, 2011; Schenker, 2015). Facebook has continued to dominate the social media landscape worldwide and its acquisition of Instagram and WhatsApp has solidified that position (Page, 2015).

Perhaps the fact that these social media applications were "free to use" and were sustained by user-generated content appealed to libertarian hopes that ordinary users would be able to steer the Internet away from its military origins and towards the interest of civil society. However, it is harder to be so optimistic today with developments such as cookies, malware, and customized news and advertising feeds, which represent the alignment of Big Tech, retailers, and any business that could benefit from knowledge about users' social networks, shopping habits, and preferences. As Big Tech works together with coders and experts in predictive analytics, vast quantities of user data can be collected then redirected back at users in ways that allow corporations to both predict and manipulate behaviour with shocking precision. Media reports about these activities, like Target's ability to predict a teenager's pregnancy before she had told her parents (Hill, 2012) and Facebook's manipulation of news feeds to experiment with users' emotional reactions (Goel, 2014), are just the tip of the iceberg when it comes to the kinds of privacy violations users consent to when they agree to PPs and ToS they have probably not read.

As a disturbing recent example, many applications now ask for permission to tap into device microphones, which allows companies to monitor users' location as well as what they are watching on television or online. Companies at the helm of this technology, like *alphonso* (https://alphonso. tv/), can now watch individuals in real time and cross-reference how behaviours across multiple platforms (television, smartphone, laptop) combine with advertising exposure to influence where individuals go, what they do, where they shop, what they talk about, and what they buy (Sulleyman, 2017). This mic-tapping technology illustrates perfectly how control operates in all times and spaces facilitating a process where "individuals become *'dividuals'* and masses, samples, data, markets, or *banks*" (Deleuze, 1992, p. 5). The potential for citizens to rein in these processes looks increasingly bleak against the monopoly control achieved by major players in Big Tech. At present nearly all major search engines and social media apps are owned by six corporations (Figure 4.1), and the most popular (Instagram, WhatsApp, YouTube) are owned by just two – Facebook and Google (Smith & Anderson, 2018).

Corporate interests are not the only reason citizens' personal information is vulnerable to surveillance. In the early 2000s, the emotional trauma and mass paranoia that swept through the United States and its allies in the wake of 9/11 converged with the introduction of new legislation such as the

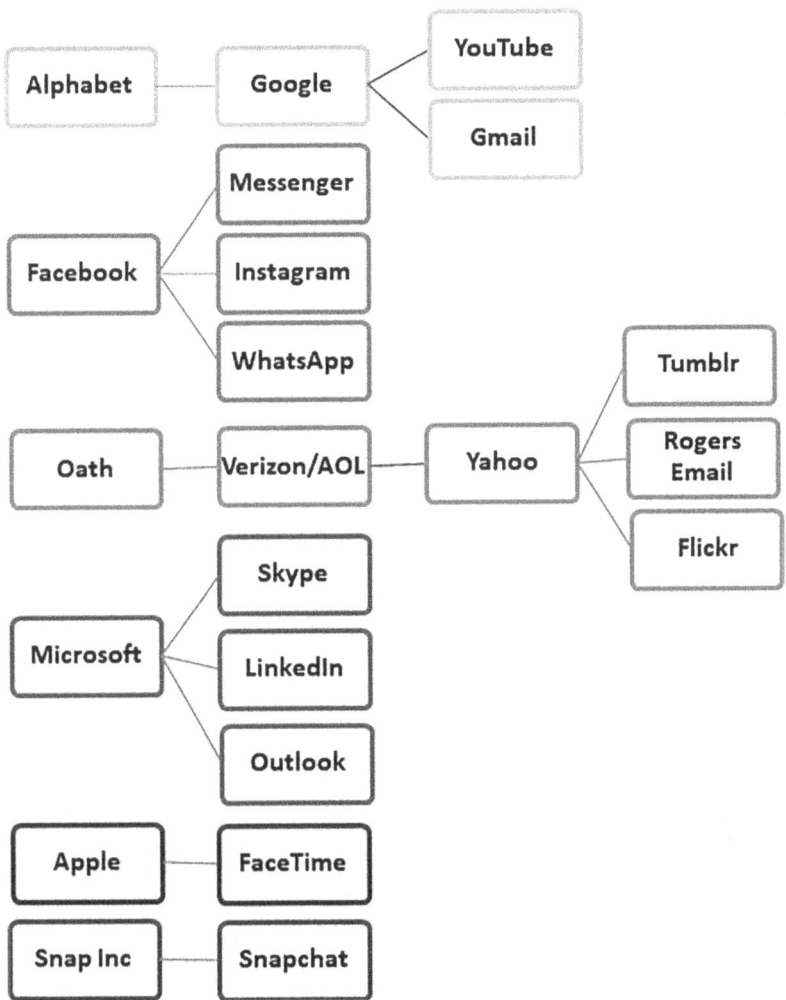

Figure 4.1 Social Media Ownership.

Patriot Act. This was a moment in history when public fear trumped civil liberties, which were all too easily curtailed in the name of circumventing terrorism. In 2013, Snowden reminded the world that more than a decade later, we are still living in a post-9/11 world. By disclosing programs like the National Security Agency (NSA)'s Planning Tool for Resource Integration, Synchronization, and Management (PRISM) he also made the world aware that companies like Facebook, Apple, Microsoft, and Google operate as arms of government agencies to intercept, deliver, and store citizens' personal communications (Gellman & Soltani, 2013). Although Snowden's claims were vigorously denied by Big Tech CEOs like Google's Larry Page and Facebook's Mark Zuckerberg in 2013 (Lee, 2013), the more recent Cambridge Analytica scandal (Greenfield, 2018) has once again cast doubt on such denials of Big Tech's collusion with the state as well as corporate integrity when it comes to protecting user privacy.

Youthful strategies for managing privacy and surveillance

A historical perspectives helps to situate the opinions and attitudes our participants held about privacy and surveillance. Born in the 1990s, these young people were undergoing the crucial stages of their social and psychological development at precisely the same moment digital devices and social media platforms were advancing and gaining mass popularity. Little wonder, then, that so many participants offered commentary on how their sense of self, their relationships, and their values were closely tied to digital technology. In many ways, the Internet, smartphones, and social media occupied a special position at the nexus of their social universe and the routine practices that defined daily life. This was very clear when Naomi (20) spoke about how each segment of her morning routine centred around checking her phone: "Waking up checking it, going to the washroom, washing my face, sitting on the couch ... it's constant for sure."

More than just habit, some even felt disoriented without technology. When asked how she would feel if she did not have access to the Internet Mary (18) responded, "I'd get bored. I'd be lost." This feeling of disorientation was also equated with social disconnection as in Louise's (16) description of life without technology: "Without it, I wouldn't be able to stay in touch with anyone, and get things out there, I would be lost without it." According to Esther (18), "it would be so weird not to have a phone right now, just weird ... I think if I didn't have a phone, I would be pretty stressed out ... I would be wondering what to do, like 'what's going on?'" Contained within these narratives is the way modulatory control works by imposing limits on worldly alternatives. As young people find themselves unable to imagine life without technology – "I cannot see how I could live without it" (Naomi, 20) – they also struggle to imagine possibilities for resisting technological control – "no one's going to take down the Internet, no one's going to stop what we're doing. We're just going to evolve it" (Lucy, 16).

For many participants, the importance of maintaining a presence online via social media generally eclipsed misgivings about one's data being harvested to satiate "FANG," a popular Wallstreet acronym for the four dominant technology companies on the Internet – Facebook, Amazon, Netflix, and Google's parent company Alphabet (Burggraf,

2017). When young people do take measures to guard their privacy, they do so mainly out of concern for either their personal safety – "I purposely don't give people my ... contact information ... my birthday is not on there" (Valentina, 16) – or fear of jeopardizing their potential with current or future employers. For example, Adam (23), like so many other participants, was careful about what he posted on his Facebook page not just because he wanted to protect his privacy, but because he wanted to create a certain impression for others: "Whenever I posted something on Facebook, I think how will this be received if someone else saw it." Sebastian (17) figured this kind of self-censoring could wait until he was older and he would "have to be like more cautious of everything that I post because eventually it's, it's going to catch up with you and if you're trying to get a job."

Interestingly, participants' comprehension of the issues surrounding online privacy varied considerably as did their practices for protecting their personal information. Many felt that they had control through "jacking up" privacy settings "as high as they can go so only my friends can see what I post" Carlos (20), and careful posting, "I don't have an issue with being public unless I'm posting personal things" (Lucy, 16). Another common practice to protect privacy was to use some applications and avoid others that were presumed to be less secure. Sebastian (17), for example, did not own a cell phone and did not want a texting app because "text goes through the company's server and then out into a regular server to another person. It's just like the amount of people who can read your texts and I don't like the thought of that." Instead, Sebastian thought his information could be kept more private if he only used Facebook's Messenger app on his iPod. Cameron (16) was convinced that "there's a lot more privacy in Snapchat than there would be on Instagram," and Raelynn (20) was careful to keep her Twitter account private because "it's open to public; everybody can see it. So I don't post very private pictures of myself on there as much as I do on Facebook and Instagram."

There were participants who felt that even with these kinds of precautions their data were never truly secure. For, Dominic (16), "privacy is only a word" and the only way to safeguard your information is "to watch out what you post." Cooper (19) spoke about "the illusion of privacy" which is a recurrent theme in the data. He went on to say: "There you are by yourself in your room and there are people who are able to come into your home at any time, with any device you use to communicate with your friends."

Even among participants who thought they could exercise control over their data and manage their impression online, there were still uncertainties. In one moment, Sebastian (17) was confident that "you can delete messages once you're done, so the messages that you send somebody will never come up again." Yet, when discussing posts he deleted from social media sites he remarked, "even if you delete them, they're not permanently deleted, some-body can find them." And he wondered, "how do you deal with that?"

The realization that adjustments to privacy settings or deleting content did little to protect personal information was also disconcerting for Sadie (18) who changed her privacy controls so that others would not be able to Google her. Yet, after doing so, she still received unsolicited emails causing her to wonder; "how were you able to e-mail me because … all my settings are set as private?" In considering this question, she concluded "even though it's private it's not as private as you would like it to be."

Contradictions were also evident in participants' beliefs about the possibility of remaining anonymous. Carson (17) was certain that he could traverse the net anonymously by disguising his Internet Protocol (IP):

> it's nearly impossible to find somebody who doesn't want to be found 'cause you can go through a proxy, or you just unplug your router, you plug it back in, you've got a new IP address. So if you don't want to be found, you won't be found.

Sebastian (17) believed protection from surveillance was an inevitable out-come of the volume of data and the layers of password protections that formed the architecture of the Internet – "when everything's on the network it's so easy to hide stuff." However, others were convinced that everything they did online could be traced back and so any attempt to disguise one's identity would be futile: "You can be anonymous, but you can also be traced back to who you are. You don't need to cover your tracks because you can't" (Dominic, 16).

There was a sense that service providers were cognizant of privacy issues and were making changes to add layers of protection: "the whole lock thing on Facebook, they didn't have a thing five years ago about that" (Raelynn, 20). However, many also expressed concern that privacy settings did not offer much control over how their information was shared by others in their networks. For example, being "tagged" by friends in their posts, or having friends share posts that are open to their "friends' friends" many of whom the original poster may not know. Another concern was that the app's distribution policies may not be transparent: "they don't ask for your consent before they change their policies or change how things will be distributed. So a photo that maybe you just wanted to send to a couple of people, it could send to everybody" (Carolyn, 17).

When speaking of privacy, participants focussed mostly on the information that they uploaded to various apps, not the background or metadata. Most of the privacy control measures they used would not stop a determined hacker, nor would they offer protection against the bulk collection and storage of user data by third parties or the government. Clearly, participants were aware of and concerned about their inability to control their data, yet, ironically, they refrained from exercising control through simple measures like reading ToS – "Who reads the information at the bottom, like the big long [terms of service]" (Bailey, 21). Participants were also generally aware of the risks that came with posting online. While some mitigated the risk by reducing use, being careful about what they posted, or by deactivating some accounts, no one removed themselves completely from social media, which was seen as a necessity, if not for social life, then certainly for school or work – "I have to … be into the digital world or else I won't succeed" (Jaxon, 20).

Youthful notions of surveillance and freedom

Edward Snowden declared in a 2013 letter that his motive for leaking secret NSA documents was to provoke a public conversation about the need to limit state surveillance and to hold powerful state officials accountable when they violate citizens' rights to privacy and protection from unnecessary search and seizure (Gellman & Markon, 2013). Only one year later, David (19) gave his own predictions about the future of state surveillance through digital technology, which he compared to Orwell's classic *1984*:

> seems like *1984*, I don't know … maybe down the road, I don't know how long, you're gonna be watched everything you do … everything you do will be watched by somebody somewhere. Like government stuff

While David was careful to point out that government and corporate surveillance had always been an issue, he believed that the future would hold something new in terms of the degree of surveillance and the range of activities that would be watched and tracked. When asked whether he was worried about this, David responded with apparent resignation: "I am and I'm not. Whatever happens, happens, I guess." He was, however, adamant that nothing could be done to curtail the growth of surveillance in the future because the scale and scope of change required would be too great: "So many people that have it and enjoy it right? It's changed that many minds." Charles (19) was also aware of, but indifferent to government surveillance. He believed Edward Snowden's actions were beneficial because now "people understand what the government really holds on them." Personally, however, his attitude was that it was "weird" how much data the government could obtain about people, but not something to worry too much about because "it's not physically harming me or anything like that."

Not everyone was as nonchalant as David and Charles. Serenity (18) stated that she was ready to leave social media if government surveillance became too extreme: "I guess [if] the government would take our stuff … they can know anything about us basically, through social media. I guess that would be my trigger and I don't want to do this anymore." Carolyn (17) spoke about how the Edward Snowden reports shattered her illusion that digital technology was creating a social utopia:

> any utopia is really a dystopia in disguise. So there's the whole thing with Edward Snowden ... saying like 'The U.S.A. is spying on what you're doing online They will spy ... and they will buy your information.' So that's kind of a way the utopia is shattered.

Some believed state surveillance was necessary. Willow (18) was concerned about terrorist groups that could use viruses to infiltrate government agencies, nuclear power plants, and personal computers – "There are terrorists groups who maybe connect on the Internet, but there are terrorist groups that work solely through the Internet too" Adeline (18) was also more concerned about safety than privacy and believed that hacker's abilities to "get into all sorts of different sites and mess stuff up" justified government surveillance – "so they can catch the online criminals."

Participants' narratives about privacy and surveillance indicate they knew they were being watched, exploited, and controlled when they used digital technologies. Sometimes they even knew *how*, but for the most part they were willing to accept this as the price of success in the information economy and as a necessary condition for social membership. Regardless of their level of awareness, however, it was rare for participants to link their concerns over personal privacy and security to broader issues of democratic freedom.

Conclusion

According to Edward Snowden, data mining technology "represents the most significant new threat to civil liberties in modern times" (Snowden as cited in Harding, 2014, para. 8). A study by the Pew Research Center indicates that only about 25% of American adults have altered their behaviour after reading about the Snowden leaks (Rainie & Madden, 2015, "Some people have changed," para. 4). The kinds of behavioural changes they made are similar to those of our participants – they stopped searching for information or visiting websites they would have before, they avoid discussing sensitive topics, or they avoid posting personal information (Rainie & Madden, 2015, "Some people have changed," para. 2). However, these kinds of behavioural modifications are exactly the opposite of what Snowden wanted to happen. As Snowden explained: "You shouldn't change your behavior because the government agency somewhere is doing the wrong thing. If we sacrifice our values because we're afraid, we don't care about those values very much" (Snowden as cited in Lu, 2015, para. 22). A few of our youth participants were critical about how modifying their behaviours or privacy settings in response to corporate or government wrongdoing is also a form of control that constrains their freedom. However, they continued to use these strategies believing they were the best and only forms of personal protection.

Deleuze (1992) warned, even before the Internet became mainstream, that a new phase of capitalism was causing a crisis such that control over individuals would be achieved by creating a world with no outside. His premonitions appear to have been realized in the inability of many participants to imagine alternatives to the ways in which digital technologies work in their present world. Most of their strategies for dealing with surveillance and privacy accept the world as it is and function to keep it intact. As participants spoke about privacy and surveillance, they also described feeling the force of modulatory controls such as the social pressures that comes with rapid technological change. This provoked anxieties around the need to compete by maintaining a positive digital footprint online and by keeping up with peers. Modulatory control was also expressed as a desire to avoid feeling lost or disoriented. It was these feelings that often overrode concerns around privacy. Participants, therefore, helped us to understand that resisting control by withdrawing from social media or refusing to own a cell phone is not just a matter of social exclusion or missed professional opportunities. In an age where one's identity and social belonging depend on hyper-visibility online, going offline or refusing to carry a smartphone comes with a very real risk of social death or, as Elena put it, becoming a "ghost." It is hardly surprising that "ghosting" someone, meaning to block them from your social media or ignore their texts, calls, and emails (https://www.urbandictionary.com/define.php?term=Ghosting), has become a severe social sanction.

The fusion of digital technology and 21st-century life helps to explain how it is possible for participants to be aware of and deeply concerned about the ways digital technologies threaten their personal privacy, safety, and freedom, yet still be ambivalent in their assessments about whether this technology is ultimately beneficial or detrimental, liberating or constraining. Nevaeh (20) sums up this ambivalence in her conclusion that "it's just how people use it, use their technology; they're either going to use it for good or use it for bad." Meanwhile, as quoted earlier, Madelyn (16) holds society accountable: "It's like we're not actually free from the technology ... I find that it's strange, like you're trying to fit in, almost, with society. But we are a society who's doing this to ourselves." These statements represent a widely held belief among participants that technology is both good and bad because, in the end, it is us.

Notes

1 Quotations have been edited for readability: expressions such as "um" and "like" have been removed, and ellipses indicate when a word, sentence, or section has been removed but the meaning of the quote has not changed.

2 All participants have been given a pseudonym.
3 Facebook, Google, Apple, Amazon, and Microsoft (Lotz, 2018).
4 Pinging initially referred to the method of locating a phone's location by identifying the cell phone tower of the phone's last received signal. Today, Apple's iOS, Google's location history, and various software applications permit more efficient and accurate means of pinging a phone's location in real time (James, n.d., para. 1–2).
5 As an interesting example, in the early 2000s, the US military began rethinking its Psychological Operations (psyops) strategy in the context of a globally networked world where "the distinction between foreign and domestic audiences becomes more a question of USG intent rather than information dissemination practice" (DoD, 2003, p. 26).
6 In the United States, this has been apparent in debates and court cases surrounding the "third party doctrine" (ACLU Staff, 2018).

References

ACLU Staff. (2017). *The right to keep personal data private: Carpenter v. U.S.* Retrieved from https://www.aclu.org/blog/privacy-technology/location-tracking/right-keep-personal-data-private-carpenter-v-us

Ahn, J. (2011). The effect of social network sites on adolescents' social and academic development: Current theories and controversies. *Journal of the American Society for Information Science & Technology, 62*(8), 1435–1445. doi:10.1002/asi.21540

Almeroth, K. C. (1999). The evolution of multicast: From the MBone. *Stardust.com*, Retrieved from https://www.cs.umd.edu/projects/syschat/multicast1.pdf

Althusser, L. (1971). *Ideology and ideological state apparatuses: Lenin and philosophy and other essays.* (B. Brewster, Trans.). London, UK: New Left Books.

Asano, E. (2017, Jan 4). How much time do people spend on social media? [infographic]. *SocialMediaToday.* Retrieved from https://www.socialmediatoday.com/news/how-much-time-do-people-spend-on-social-media-infographic/450443/

Ayres, I., & Schwartz, A. (2014). The no-reading problem in consumer contract law. *Stanford Law Review, 66*(3), 576.

Baldwin, H. (2015, Mar 23). Drilling into the value of data. *Forbes.* Retrieved from https://www.forbes.com/sites/howardbaldwin/2015/03/23/drilling-into-the-value-of-data/

Berreby, D. (2017, Mar 03). Click to agree with what? No one reads terms of service, studies confirm. *The Guardian.* Retrieved from https://www.theguardian.com/technology/2017/mar/03/terms-of-service-online-contracts-fine-print

Burggraf, H. (2017, Jun 19). New acronym added to International Investment's compendium. *International Investment.* Retrieved from http://www.international investment.net/other/new-acronym-added-international-investments-compendium/

Burston, J., Dyer-Witheford, N., & Hearn, A. (Eds.). (2010). Digital labour [Special issue]. *Ephemera, 10*(3/4), 214–539.

Chan, R. (2017, Feb 22). For IoT devices, think beyond "Wow, that's so cool": Consider the privacy you are giving up. *Inverse.* Retrieved from https://www.inverse.com/article/27389-jay-stanley-aclu-privacy-internet-of-things

Coleman, G. (2009). Code is speech: Legal tinkering, expertise, and protest among free and open source software developers. *Cultural Anthropology, 24*(3), 420–454. doi:10.1111/j.1548-1360.2009.01036.x

Dayen, D. (2017, Dec 26). Big tech: The new predatory capitalism. *The American Prospect.* Retrieved from http://prospect.org/article/big-tech-new-predatory-capitalism

Deleuze, G. (1992). Postscript on the societies of control. *October, 59* (Winter), 3–7.

Department of Defense (DoD), United States of America. (2003, Oct 03). *Information operations roadmap.* Retrieved from the National Security Archive https://nsarchive2.gwu.edu/NSAEBB/NSAEBB177/index.htm

Dyer-Witherford, N. (1999). *Cyber-Marx: Cycles and circuits of struggle in high technology capitalism.* Chicago IL: University of Illinois.

Dyer-Witherford, N. (2015). *Cyber-proletariat: Global labour in the digital vortex.* London, UK: Pluto.

Foucault, M. (1990). *The history of sexuality: An introduction (Vol.1)* (R. Hurley, Trans.). New York, NY: Vintage Books.

Foucault, M. (1995). *Discipline and punish: The birth of the prison* (2nd ed.) (A. Sheridan, Trans.). New York, NY: Vintage Books.

Fuchs, C. (2014). *Digital labour and Karl Marx.* New York, NY: Routledge.

Gellman, B., & Markon, J. (2013, Jun 10). Edward Snowden says motive behind leaks was to expose 'surveillance state'. *The Washington Post*, Retrieved from https://www.washingtonpost.com/politics/edward-snowden-says-motive-behind-leaks-was-to-expose-surveillance-state/2013/06/09/aa3f0804-d13b-11e2-a73e-826d299ff459_story.html?utm_term=.928e5974a7fe

Gellman, B., & Soltani, A. (2013, Dec 04). NSA tracking cellphone locations worldwide, Snowden documents show. *The Washington Post.* Retrieved from https://www.washingtonpost.com/world/national-security/nsa-tracking-cellphone-locations-worldwide-snowden-documents-show/2013/12/04/5492873a-5cf2-11e3-bc56-c6ca94801fac_story.html?utm_term=.92599935f472

Goel, V. (2014, Jun 30). Facebook tinkers with users' emotions in news feed experiment, stirring outcry. *New York Times.* Retrieved from https://www.nytimes.com/2014/06/30/technology/facebook-tinkers-with-users-emotions-in-news-feed-experiment-stirring-outcry.html

Greenfield, P. (2018, Mar 26). The Cambridge Analytica files: The story so far. *The Guardian.* Retrieved from https://www.theguardian.com/news/2018/mar/26/the-cambridge-analytica-files-the-story-so-far

Grossberg, L. (2010). *Cultural studies in the future tense.* Durham, NC: Duke University Press

Hall, S. (1977). Re-thinking the 'base-and-superstructure' metaphor. In J. Bloomfield (Ed.), *Papers on class, hegemony and party* (pp. 43–72). London, UK: Lawrence & Wishart.

Harding, L. (2014, Apr 08). Edward Snowden: US government spied on human rights workers. *The Guardian.* Retrieved from https://www.theguardian.com/world/2014/apr/08/edwards-snowden-us-government-spied-human-rights-workers

Hill, K. (2012, Mar 31). How Target figured out a teen girl was pregnant before her father did. *Forbes.* Retrieved from https://www.forbes.com/sites/kashmirhill/2012/02/16/how-target-figured-out-a-teen-girl-was-pregnant-before-her-father-did/#7f6a00486668

Industry Canada, Office of Consumer Affairs. (2006). *The expansion of cell phone services.* Retrieved from http://www.ic.gc.ca/eic/site/oca-bc.nsf/eng/ca02267.html

James, E. (n.d.). How to ping a cell phone location. *Techwalla.* Retrieved July 30, 2018, from https://www.techwalla.com/articles/how-to-ping-a-cell-phone-location

Jenkins, H. (2006). *Convergence culture: Where old and new media collide.* New York, NY: NYU Press.

Laboisse, P. (2017, Nov 22). Sensor technologies will drive the next digital age. *Electronic Design.* Retrieved from https://www.electronicdesign.com/analog/sensor-technologies-will-drive-next-digital-age

Lee, T. B. (2013, Jun 12). Here's everything we know about PRISM to date. *The Washington Post.* Retrieved from https://www.washingtonpost.com/news/wonk/wp/2013/06/12/heres-everything-we-know-about-prism-to-date/?utm_term=.6dd2acc67b4b

Lotz, A. (2018, Mar 23). 'Big Tech' isn't one big monopoly - it's 5 companies all in different businesses. *The Conversation.* Retrieved from https://theconversation.com/big-tech-isnt-one-big-monopoly-its-5-companies-all-in-different-businesses-92791

Lu, A. (2015, Apr 06). John Oliver and Edward Snowden's dick pics conversation has to be seen to be believed. *Bustle.* Retrieved from https://www.bustle.com/articles/74355-john-oliver-edward-snowdens-dick-pics-conversation-has-to-be-seen-to-be-believed

Meola, A. (2018, May 10). What is the Internet of Things (IoT)? Meaning & definition. *Business Insider.* Retrieved from https://www.businessinsider.com/internet-of-things-definition

Nanos, J. (2018, Jul). Every step you take. *The Boston Glove.* Retrieved from http://apps.bostonglobe.com/business/graphics/2018/07/foot-traffic/

Newman, L. H. (2017, May 02). Hundreds of apps can listen for marketing 'beacons' you can't hear. *Wired.* Retrieved from https://www.wired.com/2017/05/hundreds-apps-can-listen-beacons-cant-hear/

Obar, J. A., & Oeldorf-Hirsch, A. (2016). *The biggest lie on the Internet: Ignoring the privacy policies and terms of service policies of social networking services [DRAFT].* Paper presented at TPRC 44: The 44th Research Conference on Communication, Information and Internet Policy. doi:10.2139/ssrn.2757465

Page, V. (2015, May 18). Top companies owned by Facebook (FB). *Investopedia.* Retrieved from https://www.investopedia.com/articles/personal-finance/051815/top-11-companies-owned-facebook.asp

Rainie, L., & Madden, M. (2015, Mar 16). Americans' privacy strategies post-Snowden. *Pew Research Center.* Retrieved June 01, 2018, from http://www.pewinternet.org/2015/03/16/Americans-Privacy-Strategies-Post-Snowden/

Schenker, M. (2015, May 12). Former MySpace CEO explains why MySpace lost out to Facebook so badly. *Digital Trends.* Retrieved from https://www.digitaltrends.com/social-media/former-myspace-ceo-reveals-what-facebook-did-right-to-dominate-social-media/

Shekhtman, L. (2017, Nov 29). Phones vulnerable to location tracking even when GPS services off. Retrieved from https://www.princeton.edu/news/2017/11/29/phones-vulnerable-location-tracking-even-when-gps-services

Short, J., & Todd, S. (2017). What's your data worth? *MIT Sloan Management Review, 58*(3), 17.

Smith, A., & Anderson, M. (2018). *Social media use in 2018.* Washington, DC: Pew Internet & American Life Project. Retrieved from http://www.pewinternet.org/2018/03/01/social-media-use-in-2018/

Stanley, J. (2017, Jan 13). The privacy threat from always-on microphones like the Amazon Echo. *ACLU.* Retrieved from https://www.aclu.org/blog/privacy-technology/privacy-threat-always-microphones-amazon-echo

Sulleyman, A. (2017, May 08). Android apps are secretly tracking your location. *Independent*. Retrieved from https://www.independent.co.uk/life-style/gadgets-and-tech/news/android-apps-beacons-tracking-users-inaudible-sound-hidden-adverts-ultrasonic-audio-privacy-phones-a7723871.html

Weber, S., & Wong, R. Y. (2017). The new world of data: Four provocations on the Internet of Things. *First Monday, 22*(2). Retrieved from https://www.firstmonday.dk/ojs/index.php/fm/rt/printerFriendly/6936/5859

Williams, R. (1977). *Marxism and literature*. Oxford, UK: Oxford University Press.

Chapter 5

"It's almost like the earth stood still"

Youthful critiques of cell phones

Ron Srigley and Kate C. Tilleczek

Introduction

This chapter explores several young people's experiences of the place of cell phones in their lives through an odd and rather serendipitous exercise in which they voluntarily went without them for a period of nine days and then wrote about the experience. The results of their essays were not shocking in the sense of being unexpected. However, they were surprising in the degree to which they challenged the usual narratives about the benefits of such devices and the platforms they support and in the vividness with which they described that challenge. The following narrative describes the experiences that Ron and his students shared in living without cell phones, writing about those experiences and reflecting on that writing and its meaning.

I was teaching recently an undergraduate course in philosophy at a small Canadian university. Many students at the institution struggled academically due to limited literacy and a lack of university preparation. As troubling as this was, however, it was nothing new. I have taught dozens of courses to thousands of students over the past 20 years and it is clear that such limitations are the new normal in post-secondary education in Canada. Most professors simply expect them. But the students in this particular course were really struggling. They were not merely having trouble understanding the books in the course; they were finding them difficult simply to read. I will never forget an exchange with a student that made the whole situation stunningly clear. Frustrated by my students' silence during a discussion of Plato's *Republic*, I asked them to explain what it was that they did not understand. Long silence. Then one brave woman raised her hand and said, "We don't understand what it says, sir. We don't understand the words." I looked around the class and saw all these guileless heads nodding pensively in agreement.

It was quite a moment. Faced with an unresponsive class, most professors feel they have failed. They think that if students are "not getting it," or are not getting into it, it is the professor's fault. They are not relevant. They are not clear. Or worse, they are boring. And sometimes they are right. But this situation was more complicated. There was very little teaching and very little learning going on in that classroom. But not necessarily because no one cared if there was or was not. The students were not learning because they could not, and I was not teaching because I had not yet understood why.

My students, like all others in Canada, had been attending schools for some 13 years of their lives, yet they couldn't read a book like the *Republic*, let alone be moved, angered, or inspired by its argument or beauty. That was new. My father grew up in the 1950s in a poor, working-class neighbourhood in an industrial Canadian city. He did not go to the "right" school, and he and his classmates could only dream of attending university or college. The furthest he got was grade 10, when he was forced to leave school to support his family. Yet, by the age of 15, and despite the myriad social and economic limitations of his life, limitations that now set all pious souls aflutter with the need to accommodate, his school insisted that he read books like Homer's *Iliad* and *Odyssey*, works arguably as strange or stranger than the one I had assigned my first-year university students. And yet he did so and even 50 years later could still recall the experience. What had happened to our schools and universities?

The destruction of Canadian secondary and post-secondary education has been going on for some 70 years now, so the damage to my students was not likely to be remediated during a four-month introductory university course. But since one has to start somewhere, I began to develop a variety of what I considered legitimate accommodations for them – that is, accommodations designed to help them improve, not to affirm them in their inability. I slowed down the pace of the course, I prepared study notes for the more challenging passages, and I reformulated the analysis I was developing in simpler terms.

When the fruits of our collective labour arrived in the form of the students' midterm tests, I was discouraged to the point of despair. All that effort for nothing. Nothing. Even the more capable students showed no real sign of having understood either the texts or the lectures. When it gets that bad, one really wonders what the point of the whole business is, apart from professors' need for a vocation and students' need for a credential – in other words, apart from our collective need to keep the big, expensive educational machine running. But my despair notwithstanding, there was still the practical matter of how to assess their work. Could I really fail half the class? I might be able to get away with that if I were teaching organic chemistry. But in an elective (read, "service") course like philosophy? I would have every administrator in the university, plus a few invented especially for the occasion, come down on me.

In order to solve this assessment impasse (my unwillingness to give grades for nothing, combined with the impossibility of failing half the students), and based on a hunch I had been having about the connection between chronic in-class cell phone use and academic performance, I offered my students a compromise: Give me your cell phones for the next nine days, then write an essay about the experience of living without it. This will give you a chance to earn additional grades to bring yours into a normal range.[1]

Twelve out of the 35 students in the class chose to participate. I told them three different grades were possible: Excellent (5/5), Acceptable (3/5), and Submitted (1/5). I also strongly encouraged them to say what they really thought and experienced. I assured them, for instance, that there were no "right" or "wrong" answers, just better and worse ones. And I made it clear that even a hint of pandering to what they assumed to be "the professor's point of view" would result in lower grades, not higher ones. In other words, I told them they were to act like the university students of old and write freely and frankly about what they felt and thought without fear of reprisal, if not of criticism. I should also mention that digital technology was not a direct object of study in the course, so students were not responding to my own analysis of that phenomenon, though they would certainly have heard me make various extemporized comments about it at various points in the course.

In the event, all students received five out of five, not because their efforts were equally successful but because all of them spoke with an honesty, interest, and at times eloquence about their experience that was infinitely better than any other written work they produced in the course. That, too, was quite a shock – a welcome one this time. For the first time in years I had the sense that there was something profound there to work with, if only I could encourage them to put their cell phones away long enough to get them to explore and express it. That was my end game in the course. But what they actually said about their cell phone-less lives was extraordinary because of both how reflective it was and how radically it departed from standard education and tech-industry narratives about how wonderful this technology is.

The following pages are an analysis of what they said. Pseudonyms have been employed throughout to protect identities and ensure confidentiality. We have organized the analysis according to the broad themes that emerged from their essays. As readers will note, there was an extraordinary level of agreement, if not of emphasis, among these students. What is perhaps most striking is how emotionally, socially, and intellectually deleterious they came to think their cell phones were after living without them for just over a week. What they said was virtually the opposite of the advertisements we are accustomed to hearing from educational authorities and technology advocates. The themes and expressions also resonate with those of the young people in other chapters in this book. Though the modes of speaking to these young people differed, there is a remarkable agreement in their thinking about the meaning of technology in their lives.

What to make of this thing? A general assessment

The usual industry and education narrative about the advantages of cell phones, social media, and digital technology generally is well known. It is claimed that such devices build community; foster communication;

increase efficiency; and, as a result, improve the lives of their users. Mark Zuckerberg's recent (2017) reformulation of Facebook's mission statement is typical of this narrative: The company's ambition is to "give people the power to build community and bring the world closer together." He then went on to give some content to this ambition: "ending poverty, curing disease, stopping climate change, spreading freedom and tolerance, stopping terrorism." No mean feats, those. Certainly not ones that a "single group or even country can do" on its own. Nonetheless, Zuckerberg exhorted us to strive to do so by building countries and groups and, one assumes, companies too, willing to take on "these big meaningful efforts" (Zuckerberg, 2017).

The immediate, ostensible reason for Zuckerberg's reformulation, complete with its modern pieties, was that foreign political operators had been caught using Facebook's platform for precisely the opposite reasons: In other words, to spread division, undermine communities, and encourage hostility and mistrust (Rosenberg & Frenkel, 2018). However, an even more troubling critique had already begun to emerge, one that went beyond the obvious political conspiracies to explore the design and ambitions of the platform itself. For instance, high-profile defectors from the industry began to explain the ways in which cell phones and the social media platforms they support are not merely open to political abuse, but are designed knowingly so as to be harmful to users personally and to societies generally in pursuit of a business model that actually depends on that harm for its profitability (Kent, Sottile, Goss, & Newcomb, 2018). Roger McNamee, former advisor to Mark Zuckerberg, commented that

> All the content [you see on Facebook] is stuff that you like, right? It's what they think you like. But what it really is, is stuff that serves their business model and their profits ... And making you angry, making you afraid, is really good for Facebook's business. It is not good for America. It's not good for the users of Facebook. (Kent et al., 2018, para. 16)

Tristan Harris, a former design ethicist at Google, explained both the source and purpose of that anger and fear: "What people don't know about or see about Facebook is that polarization is built in to the business model ... Polarization is profitable" (Kent et al., 2018, para. 12). Dividing people, making them fear one another and compete with one another, may undermine a society's well-being, but it also provokes the very sort of feverish online activity from which technology companies profit most.

British journalist and novelist John Lanchester explored this model in his groundbreaking study of Facebook published in 2017 in *London Review of Books*. In "You Are the Product," Lanchester cites an article in the *American Journal of Epidemiology* entitled "Association of Facebook Use with Compromised Well-Being: A Longitudinal Study." According to Lanchester, the study "found quite simply that the more people use Facebook, the more unhappy they are." Even more, the study suggested "the positive effect of real-world interactions, which enhance well-being, was accurately paralleled by the 'negative associations of Facebook use.'" In effect, Lanchester told us, "people were swapping real relationships which made them feel good for time on Facebook which made them feel bad." He summed up the consequences of chronic use of such platforms: "there is a lot of research showing that Facebook makes people feel like shit" (Lanchester, 2017, para. 43).

These students could not agree with Lanchester more. Their essays not only confirm his analysis but also extend it by clarifying just what "feeling like shit" looks like on the ground and how its consequences stack up against the standard narrative of people like Zuckerberg. It is about time we started listen to these young people. They are the real canaries in the coal mine of our brave new technological and digital world.

"You must be weird or something": what the students said

Six distinct though overlapping themes emerged from the student essays: human relationships, freedom, productivity and focus, morality and engagement, parents, and safety. None of these themes were assigned or even suggested to them. They emerged spontaneously from the students' own thinking about their experiences. And there was a remarkable consistency in their thinking, despite the different emphases on and assessments of cell phones. For instance, the majority claimed that their cell phones were having deleterious effects on their human relationships, their productivity and focus, their freedom, and even their morality, and at least half of them thought that cell phones had cut them off from the "real" world in preference of an artificial technological reality. One student did write favourably and even apologetically about his cell phone and its uses. Yet even he qualified that approbation by indicating that cell phones could and often did have the types of consequences identified by the other students. But for him, that was the fault not of the cell phone, which is merely a tool, but of the person using it.

One final theme stands out for comment, though it was mentioned explicitly by only one student: writing. My experience in this class was similar to my experience in virtually all of my classes – students have a great deal of difficulty expressing themselves in prose. One student in the group acknowledged this fact and offered a possible explanation: cell phones and the laziness they encourage. The irony of her comment, however, was that in the very essay in which she made it, her writing skills demonstrated a notable improvement. By improvement I do not mean her essay was grammatically sound or stylistically refined or sophisticated. On those fronts this student's work remained, like most of her colleagues', moderately to seriously flawed. No, it was not technical proficiency that had changed in these essays but rather their spiritedness or engagement with the subject. Most of their writing to this point in the course had been bland and perfunctory at best. But this assignment was different. You could hear their voices, and you could sense what they felt about what they had experienced. Indeed, you could sense that they were trying to find the words to express their experiences, and not just pumping out academic jargon. What they were saying somehow mattered to them and you could see it in their writing.

All the themes and features of their essays require further discussion. We will begin with human relationships.

Human relationships

> "Believe it or not I had to walk up to a stranger and ask what time it was. It honestly took me a lot of guts and confidence to ask someone" (Janet).

Janet's comment was one I'd read more than once in these essays. It is also one I hear frequently from other young people today. Of course, it is not unusual for youths to feel uncertain and awkward in public. However, what Janet was saying was not merely that talking to people she doesn't know is difficult, but that it's strange or abnormal, even if the request is for something as simple as the time. Cell phones seem to encourage this strangeness in several ways: "Why do you need to ask me the time? Everyone has a cell-phone. You must be weird or something." This was Janet's concern – that people would think she was strange or lying on the assumption that everyone alive has a cell phone. Emily went even further. Simply walking "by strangers in the hallway or when I passed them on the street" caused almost everyone to take "out their phone right before I could gain eye contact with them."

To these young people, direct, unmediated human contact was experienced as ill-mannered at best and at worst as strange. Cell phones served as a type of social protection from such "live" or "real" situations. If you simply did not want to, or if you did not know how to cope with some person or situation, all you had to do was take out your phone and look down. James:

> One of the worst and most common things people do now a days [sic] is pull out their cell phone and use it while in a face to face conversation. This action is very rude and unacceptable, but yet again, I find myself guilty of this sometimes because it is the norm.

Stewart described what it was like to spend a week without this protection:

> Conversations I didn't want to be in, or situations that were uncomfortable, I was forced to endure and interact in because I didn't have my little hand-held device to shield me from any awkwardness that happens in day to day [sic] life.

And from her cell phone-less perch at a party, Emily noticed that

> a lot of people used their cell phones when they felt they were in an awkward situation, for an example being at a party while no one was speaking to them. I noticed it a lot whenever my group of friends stop speaking for a minute, they all took out their cell phones.

The price of this protection, however, was the loss of human relationships, a consequence that almost all of the students identified and lamented. During the week without his phone, James found himself forced "to look [people] in the eye and [be] engaged in the conversation. This proves the fact that these things can be very negative when over used and when people depend on them too much." For him, the cell phone had a narcotic effect that caused a loss of "people skills" and, interestingly, of "intelligence": "We have become addicted to these devices and they are throwing our intelligence and people skills down the drain." For James, intelligence derives in part from human relations. Lose those and you lose your mind, as it were. And Stewart put a moral spin on the experience. "Being forced to have [real relations with people] obviously made me a better person because each time it happened I learned how to deal with the situation better, other than sticking my face in a phone." Of the 12 students, fully 10 of them said that their phones were compromising their ability to have "real" relationships with people and that spending a week without them made them keenly aware of this fact.

Freedom

> "I have to admit, it was pretty nice without the phone all week. Didn't have to hear the fucking thing ring or vibrate once, and didn't feel bad not answering phone calls because there were none to ignore" (Peter).

The reverse side of this absence of real human relations was an awareness of a proliferation of meaningless "communication" with people encouraged by their cell phones. Not surprisingly, virtually all students admitted that ease of communication – with employers, friends, parents – was one of the genuine benefits of their phones. However, eight out of twelve said they were genuinely relieved not to have to answer the usual flood of texts, messages, and social media posts. Indeed, the language they used to describe their freedom indicated that they experienced this activity almost as a type of harassment. "It felt so free without one and it was nice knowing no one could bother me when I didn't want to be bothered," wrote Peter. Emily said she found herself

> sleeping more peacefully after the first two nights of attempting to sleep right away when the lights got shut off. I didn't frantically look for my cell phone in my bed that managed to disappear through the night under the sheets (yes I sleep with my cell phone).

Elliott wrote that "The thing that changed most during my week without a cell phone was that I had a lot more free time and I could concentrate longer on different things." And Edward admitted that what he

> liked the most about not having a phone was the fact that [he] didn't have to be constantly talking to people. It was nice to be able to do something without thinking about when [his] phone is going to buzz again.

Several students went even further and claimed that "communication" with others was in fact easier and more efficient *without their phones.*

> Actually I got things done much quicker without the cell because instead of waiting for a response from someone (that you don't even know if they read your message or not) you just called them [from a land line], either got an answer or didn't, and moved on to the next thing. (Stewart)

One of the primary benefits of cell phones and of the social media platforms they support is said to be that they facilitate communication and bring people "closer together." In fact, these students experienced the communication provided by these devices almost as a distracting nuisance and as something that in part deprived them of genuine human relationships.

Productivity and focus

How about productivity? One thing that seems to be incontrovertible is the technologists' assertion that their instruments and platforms make us more productive. What have the students to say about this?

If we are speaking about speed, most students would agree. The one thing cell phones and computers are is fast. But for them that speed didn't translate into greater or better productivity. In fact, it had the opposite effect. Elliott claimed that without his phone he

> could get everything done faster, for instance writing a paper and not having a phone boosted productivity at least twice as much. You are concentrated on one task and not worrying about anything else. Studying for a test was much easier as well because I was not distracted by the phone at all.

Stewart found he

> could sit down and actually focus on writing a paper. Because I was able to give it 100% of my attention, not only was the final product better then [sic] it would have been, I was also able to complete it much quicker.

And James said that

> Before, with the distraction of the cell phone present, I found that when doing school work I would easily become unfocused. In fact, I found myself taking breaks often to play around with my cellphone. When that distraction was removed I discovered that I got much more work done in a shorter period of time.

Even Janet, who missed her cell phone more than most, admitted that

> One positive thing that came out of not having a cellphone was that I found myself more productive and I was more apt to pay attention in class. I didn't get distracted as much, always looking at my phone and texting people.

For her the only "downside" was that "Class did go by a lot slower and some classes felt forever without my cellphone."

The cell phone and its social media platforms did not make these students more productive; it distracted them and made them considerably less productive, and in the process diminished the quality of their work. This should not surprise anyone given what we now know about these technologies. Their primary aim *is* to distract people, to move them around sites, to get them to click as much as possible, and then to compile the resulting data and monetize it. Facebook, for instance, has no interest in people using its platform modestly and judiciously (Lanchester, 2017). Excess is where the money is, and the design operates accordingly. Yet many of our educational experts continue to advocate including more of this technology in schools and universities (Contact North, n.d.).

Morality and engagement

> "Having a cell phone has actually affected my personal code of morals and this scares me" (Gina).

Morality is a tricky subject in liberal regimes and education systems for the simple reason that in such regimes and systems morality is considered a matter of private opinion and so everyone is permitted to have their own (Mouffe, 1991).[2] What this means is that serious public discussion about moral problems tends to be as brief as it is empty and is usually confined to the matter of calculating personal interests. Nonetheless, even we good liberals, if you push us hard enough, will begin to talk about things like fairness and justice and to express normal human concerns about situations that fail to achieve them (Mouffe, 1991).[3] My students were no exception. Ask them a question about the "best regime," and they would invariably answer, "that depends on a person's personal opinion." But ask them about the effects their cell phones are having on their lives, and they were quick to tell you how morally compromised they feel and how upset they are about it.

Gina was a case in point:

> Since I started using a cell phone three years ago I have noticed that I have become increasingly attached to it. I regret to admit that I have texted in class this year, something I swore to myself in high school that I would never do ... I am disappointed in myself now that I see how much I have come to depend on technology in the last few years. I start to wonder if it has affected who I am as a person, and then I remember that it already has.

And James, though he claimed that we must continue to develop our technology because "many lives will depend on it," said that "what many people forget is that it is vital for us not to lose our fundamental values along the way."

When it came to explaining the content of those values, students were less clear. But they were by no means silent. They were worried about their addiction to these devices, not just abstractly, the way they are supposed to be worried about being addicted to drugs, but substantively. They had a sense of what that addiction was depriving them of: a meaningful relationship to the world. This was, of course, a philosophy course, one in which there were discussions of Plato's image of the cave in book 7 of the *Republic*, so it is not surprising that the students would express themselves in this way. After all, Plato's image depicts people imprisoned in a shadowy dream world of illusions and explores the manner of their release and return the world beyond the cave. The parallels to modern screen technologies should be obvious.

Nonetheless, I had the impression that the students were also attempting to describe actual experiences they'd had during the time without their phones rather than merely repeating tropes from the course readings or lectures. Listen, for instance, to James as he described his experience:

> It is almost like the earth stood still and I actually looked around and cared about current events; I was not blinded by media. I know that sounds extravagant but it is true. This experiment has made many things clear to me and one thing is for sure, I am going to cut back the time I am on my cell phone substantially.

Not only did James begin to see things; he also began to care about them. The experience of seeing something as it is was, for him, also moral experience.

Or consider Emily's remark that cell phones "take away freedom, [they take] away our mind and knowledge to the outside world and [they take] away our confidence. It's interesting to say that [they take] away our knowledge to [sic] the outside world." For Emily, as for James, when you lose contact with life, you lose the source of your intelligence, too. And to explain the consequences of that loss of contact and intelligence, she appealed to George Orwell's *1984*:

> I'm a huge *1984* fan and after reading that book two years ago I relate it to everything in our everyday life. What was predicted is pretty

accurate, which is scary because we are all convinced we have freedom that is given to us in short supply, and cell phones are a huge part of 'Big Brother.' We are told once that a cell phone benefits us by quickly allowing us to communicate with one another, however we are not told about the consequences that will come within a short amount of years.

Far from liberating people for real community, cell phones imprison us in a stifling technological regime that kills freedom, hampers intelligence, and makes life empty.

And then there was Stewart, who began to see how things "really work" once he was without his phone:

> One big thing I picked up on while doing this assignment is how much more engaged I was in the world around me all the time. I was always paying attention to what was going on, not just tuning things out because I had a message that had to be replied to right that minute. And you know what I notice when I was ever so engaged in the world around me? I noticed that the majority of people were disengaged, on a computer or laptop. It's pretty ridiculous, there is all this potential for conversation, interaction and learning from one another but we're too distracted by the screens that are everywhere around us to partake in the real events around us.

These students were not naïve at all; nor were they passive. Given only a small bit of freedom from the usual technological constraints of their lives, they began to see the situation very clearly and even started to resist it. The purveyors of cell phones and social media platforms have made it amply clear by their actions, if not by their words, that they wish to mediate everything we experience. The Gospel of John: "No one comes to the father but through me." The Gospel of Facebook: "no one comes to the world, but through Mark Zuckerberg's platform." Technology companies have sought aggressively to extend the scope of their influence to virtually every corner of the globe and of human experience. Their ubiquity alone makes it seem that any resistance to them is futile. Yet even after only a short stint without their phones, many of my students recognized their predicament and expressed genuine disaffection with it. It strikes me that as educators and as those who care for the young, it's up to us to ensure that students have more such experiences, not fewer, if they are ever to be able to make choices that are genuinely their own.

Parents

"My mom thought it was great" (James)
"It was extremely stressful for my mom" (Janet)

When it came to the matter of families, there were several contrasting and crosscutting views presented. Take parents, for example. At least three students indicated that their parents were quite pleased with their new, cell phone-less selves. One parent even proposed to join in the experiment, making it a family endeavour. These parents said that their children were more present and engaged in the absence of their phones, rather than being distracted and distant. James's comment was typical: "My mom thought it was great that I did not have my phone because I paid more attention to her while she was talking."

Other parents, however, were concerned about safety and the deprivation of communication. They had become accustomed to having that immediate connection with their children and did not find its absence easy. And several students also felt the same way, especially those who were out of province. Emily: "I felt like I was craving some interaction from a family member. Either to keep my ass in line with the upcoming exams, or to simply let me know someone is supporting me." And Janet admitted that "The most difficult thing was defiantly [sic] not being able to talk to my mom or being able to communicate with anyone on demand or at that present moment. It was extremely stressful for my mom."

There are of course many different ways for young people to relate to their parents. But it did seem to us that as a rule this group of university students was more frequently in contact with their parents and more dependent on them than have been previous generations of students in this country. Whether this contact and dependency were caused by cell phone technology – we are in touch more now because we can be – or the reverse is difficult to say. One thing is clear, however: They are now mutually reinforcing. Only one student, after having described his parents' concern about being unable to contact him while away on a holiday, said, "They can go fly a kite." For the rest, the parents' concern seemed legitimate and in some cases was shared.

Safety

This raises the matter of safety. Parents were concerned about their children's safety and what might happen if there were an emergency and they were unable to be in contact with one another. Several students echoed this type of concern. Janet said that "Having a cellphone makes me feel secure in a way. So having that taken away from me changed my life a little. I was scared that something serious might happen during the week of not having a cellphone." She went on to say something even more revealing:

Another thing I didn't like about not having a cellphone that made me kind of scared at times was if someone were to attack me or kidnap me or some sort of action along those line or maybe even if I witnessed a crime take place, or I needed to call an ambulance, I really wouldn't be in any position to get help for myself or anyone else if I was by myself because I wouldn't have any sort of communication to contact emergency services.

For this student and several others there was the sense that the world is a very dangerous place and that cell phones are necessary to combat that danger. This was interesting because the place where these students live has one of the lowest crime rates in the world and has almost no violent crime of any kind.[4] It should also be noted that even students from bigger urban centres were unlikely to encounter dangers of the types they feared because of their social position and class. Yet the fear of these things was very present for them and their cell phones were understood by them to be protection against it.

It seems to us that this fear is not precisely the type that Roger McNamee discussed (Kent et al., 2018). The fear he was concerned with is a sort of general fear of "the other" exacerbated by the polarization of sides that social media platforms encourage. The fear these students were talking about was more direct – fear of crime, accident, loss, etc. Nonetheless, there seems to be a family resemblance here. For there was in these students' essays, running beneath even the critiques of cell phones and the awareness of their existentially harmful character, a kind of uneasiness about being without them. Perhaps these young people, even the most prescient among them, were already on their way to the type of life being prepared for them by American technology giants like Apple, Facebook, Google, and Microsoft – a life made fearful and protected from that fear by the same device.

Conclusion

According to these students, cell phones are not the useful, productivity- and communication-encouraging devices we are accustomed to thinking of them as. In fact, if these students are to be believed, cell phones and their social media platforms frequently encourage just the opposite of these things. Rather than causing conversation (not communication) to flourish, they make it something onerous, difficult, and strange; rather than building community, they divide us; rather than making us intelligent and engaged, they dumb us down by cutting us off from the things we need most for our intelligence to grow.

So why did they keep using them, as virtually all of them said they would? Two things stood out in their essays. First, their underlying fear. It was just better to have them there, just in case. One never knows what might happen. Second, they understood in their bones that the technocratic world *we* have built for them is driven by these technologies, and therefore not to have them would seriously compromise their ability to gain a foothold in that world. They were not fools. They knew what their own lives were like. If we hold together everything they say, it is clear that they are caught in a trap – one that forces them to do harm to themselves, while we and the world around them quietly assure them that there is no other way. It is about time we started helping them look for other ways out of the trap. We are the adults in the room.

Notes

1 I've appended the actual assignment instructions at the conclusion of this chapter.
2   While liberalism did certainly contribute to the formulation of the idea of a universal citizenship, based on the assertion that all individuals are born free and equal, it also reduced citizenship to a mere legal status, indicating the possession of rights that the individual holds against the state. The way those rights are exercised is irrelevant as long as their holders do not break the law or interfere with the rights of others. Social cooperation aims only at enhancing our productive capacities and facilitating the attainment of each person's individual prosperity. Ideas of public-spiritedness, civic activity, and political participation in a community of equals are alien to most liberal thinkers

(Mouffe, 1991, p. 73)

3   To ensure our own liberty and avoid the servitude that would render its exercise impossible, we must cultivate civic virtues and devote ourselves to the common good. The idea of a common good above our private interest is a necessary condition for enjoying individual liberty

(Mouffe, 1991, p. 73)

4 To identify the location of this university might compromise the identities of the students; therefore this identity has been omitted.

References

Contact North. (n.d.) How to use technology effectively. Retrieved from https://teachonline.ca/tools-trends/how-use-technology-effectively

Kent, J. L., Sottile, C., Goss, E., & Newcomb, A. (2018, Jan 16). Facebook is a 'living, breathing crime scene' says one former tech insider. *NBC News*. Retrieved from https://www.nbcnews.com/tech/tech-news/facebook-living-breathing-crime-scene-says-one-former-manager-n837991

Lanchester, J. (2017). You are the product. *London Review of Books, 39*(16), 3–10. Retrieved from https://www.lrb.co.uk/v39/n16/john-lanchester/you-are-the-product

Mouffe, C. (1991). Democratic citizenship and the political community. In Miami Theory Collective, & G. Van Den Abbeele (Eds.), *Community at loose ends* (pp. 70–82). Oxford, MN: University of Minnesota Press.

Rosenberg, M., & Frenkel, S. (2018). Facebook's role in data misuse sets off storms on two continents. *The New York Times*. Retrieved from https://www.nytimes.com/2018/03/18/us/cambridge-analytica-facebook-privacy-data.html

Zuckerberg, M. (2017, Jun 22). Bringing the world closer together. Retrieved from https://www.facebook.com/zuck/posts/10154944663901634

Appendix: The Assignment

Bonus Assignment: Philosophy Course X

Value: Potential of 5% added to Final Grade

A. Assignment: Students agree to leave their cell phones with the professor for one week. Phones will be locked in the professor's office. Students will then write about the experience of living without their cell phone. Possible questions to be explored are:

1   What was most difficult about living without a cell phone? What was easiest?
2   What particular function of the cell phone did you miss the most?
3   What changed in your life during the week without your cell phone? What did not change?
4   How did other people react to your not having a phone (family, friends, professors)?
5   Did you like being without a cell phone? Do you not like it? Why?
6   What, if anything, did you learn about yourself that you didn't already know as a result of being without your cell phone?

B. Grades: There will be three grades possible:

5 Excellent
3 Acceptable
1 Submitted

Chapter 6

Digital capital by/for youth?

Kate C. Tilleczek and Jonah R. Rimer

Introduction

The digital age has been posited to save youth from the burdens of late modernity characterized by weakening social networks that once supported them (Furlong, 2012). This is, after all, an age that offers them the tools for eternal connection with emerging social networks that can be cashed out for personal and professional advancement. However, the lives of youth today are also confounded by growing inequalities that place them at increasing disadvantage relative to adults such as high levels of precarious labour and student debt (OECD, 2011; Statistics Canada, 2018; UNICEF, 2016; Wilkinson & Pickett, 2009). Young people are also living in a particularly individualistic time, such that problems are seen as outcomes of individual failings solved only through personal action and acquisition of individual skills (Furlong & Cartmel, 2007). This crowning feature of modernity, namely the epistemological fallacy, suggests that intensification of the optics of individual control obscures social and political relations that govern people and institutions. Individual identities and actions therefore appear as the paramount modes of understanding and problem-solving, rather than fuller analyses of social, economic, political, and networked relations. It is worth citing the original idea before returning to its elements in our analysis:

> Life in high modernity revolves around an epistemological fallacy in which feelings of separation from collectivity represents part of a long term historical process which is closely associated with subjective perceptions of risk and uncertainty. Individuals are forced to negotiate a set of risks that impinge on all aspects of their daily lives, yet the intensification of individualism means that crises are perceived as individual shortcomings ... In this context, we have seen that some of the problems faced by young people in modern societies stem from an attempt to negotiate difficulties on an individual level. (Furlong & Cartmel, 2007, p. 144)

C. Wright Mills and Dorothy Smith have also convincingly argued the importance of analyzing the ruling and political relations by which young lives are organized in the digital age (Tilleczek, 2011, 2014). Thus, the concept and practices of digital capital require analysis of technological worldviews, including the ambitions of technology companies. Our model has emerged to demonstrate how young people have waded into the digital age in hopes of leveraging digital capital while encountering further individualization, inequality, and loss of human connection.

How are youth positioned in the digital age? Variably, as both experts and novices. They are contained within the "glass cage" (Carr, 2014) of automation and mobile phones but are simultaneously enacting new forms of vagabondage across digital networks. They are both digital activists and highly prized consumers. It is well regarded that social media can be helpful for organizing activism, rallies, affinity groups, and political protests (Vaidhyanathan, 2018), such as the recent *March for Our Lives* (https://marchforourlives.com/) in the USA, organized by the young survivors of the Parkland Florida school massacre.[1] However, it is not a given that potential gains of digital media are embraced or experienced by youth. They can feel used, addicted, distracted, and unable to cash out technology against the precarious labour and education they encounter (Chapter 3; Vaidhyanathan, 2018). The following youth participants direct us to such tensions and complexities which we interrogate in this chapter:

> Social media is zoning in on a certain age group in particular … I think that is the way they'll make the most money is by selling the most vulnerable people the coolest and most trendiest [sic] things that are cool and up to date. It is stupid … but every single person I know caves into it. And I know that even older generations are still caving into it so they're doing something right in making an addictive product and a sellable product for everyone. (Naomi, 20)
>
> I guess like for a country itself and all the economics it [digital media] will help, because they could sell it to the other countries. But just talking about one person? I don't think it will affect [us] too much. (Piper, 17)
>
> LUCAS: Did you ever see the Wolf of Wall Street? *"Sell me this pen"*. That is the view. You have to do things a certain way to become as wealthy as the social media [company].
>
> AIDEN: The company gets the fortune and they are basically getting the fortune off of you and they are selling you the fame. [Interviewer: But is it fame really? Are you famous?] No, fame as in reputation-wise, to look better amongst my peers. In the case of digital technology, people make it, they design it and then they try to sell it to people, right? (Lucas, 18 and Aiden, 19)

What is digital capital?

Furlong began considering digital capital as a social feature closely related to inequality in late modernity. His plenary address at the *Australian Sociological Association* conference (Furlong, 2011) posited that access, competence, skill, and networks are the most important features of capital that young people might gain from the digital age. However, he also cautioned that digital capital includes the possibility that connectivity has come to entrench social inequalities and social reproduction and leads to exclusion; ideas echoed by others (cf. Livingstone et al., 2017; Livingstone, Mascheroni, & Staksrud, 2018; Tilleczek & Srigley, 2017).

In regard to access to digital technologies, Furlong (2011) suggested that the idea of the "digital divide" is too simplistic to be useful. Whilst it does highlight gaps between those who have access and those who do not, the ubiquity of access for youth in the Global North is rendering the idea to be ill conceived; it is not simply getting onto the Internet that matters but rather what one is able to do once online. Riley (2018) went further in showing how former Google and Facebook employees, through their *Truth about Tech* initiative, address the depth of addiction that was built into the very designs of digital media as they now abashedly lobby for products that are less addictive. She argues that the digital divide is actually inverse, with the most marginal youth using the most digital media. The known negative effects of screen time are not being communicated to all youth as Big Tech[2] has profit to gain from inundating young people, schools, and families with gadgets and screens in ways that Naomi (earlier) recognizes as "selling the most vulnerable people the coolest things." Therefore, access is an important but contested entry point to digital networks; access does not magically lead to unvarnished good or capital, does not produce positive results for everyone, and does not influence all young people in the same ways (Tilleczek & Srigley, 2017). Knowledge and skill in the digital age must include an understanding of who designs and sells digital tools, for what purposes, to what effect, and that too much screen time is potentially harmful for well-being (see Chapter 3).

Furlong's additional concepts of technical competence and navigational skill move us still further into the concept of digital capital by/for youth:

> If access alone doesn't provide much of a handle on inequalities, we might want to try and capture skill or competence. In this context, the OECD refers to a "second digital divide" which relates to the possession of skills that allow individuals to benefit from ICT. Of course, the ability to benefit cannot be adequately represented in terms of the "haves" and the "have-nots": it takes the form of a continuum stretching from those with a total lack of familiarity with ICT to those with higher-level skills. (Furlong, 2011, p. 3)

Here he captured two important points for the study of youth and digital capital. First, that competence is a resource, defined in terms of literacy and proficiency. Competence is about how well youth can (or cannot) navigate and analyze the information found online. It is about the abilities to procure and use apps, websites, and social media in ways that are personally beneficial and self-promoting through networking, selling, and marketing oneself. Second, Furlong noted that there is a range of ability and thus a crucial opening for important, new forms of education of/with youth. He linked these kinds of literacy to one's ability to use digital tools for social and economic advantages, insofar as one is able. At least, said Furlong, we are to recognize the *potential* for young people to do so. However, youth also derive sets of *navigational skills* that are not simply about technical competence but also about forms of *life management* whereby they find ways to take advantage of digital connectivity to manage the fluid and shifting complexities of the digital age. What happens when forms of life are altered by the digital age and its tools? How, for instance, do youth encounter and navigate new levels of precariousness in labour markets and educational systems to better their life chances? Can digital tools assist with navigation of structural barriers that have been created by these very tools in the first place? This is the conundrum of digital capital by/for youth. Similar to Furlong, we continue to question what youth must know in order to actively use, buy in to, avoid, resist, or manage the tools and networks to which they have been granted access. Furlong (2011) stated that this will become "about the capacity to act as a digitally equipped explorer able to use new technologies as a navigational aide to explore unfamiliar territory in ways that Beck refers to the entrepreneurship of the self" (p. 5).

Indeed, cultural capital theory provides distinctions about self-entrepreneurship by illustrating the important *interplay* between individual and structural factors. This interplay is potentially productive space(s) where youth animate digital tools, networks, and structures for self and societal betterment. Bourdieu (1986) described social capital as "the aggregate of the actual or potential resources which are linked to ... a durable network of ... institutionalized relationships" (p. 248). In this way, young people encounter people and relationships that provide access to the Internet and digital tools. Amount, type, and quality of access are then navigated and managed in this productive interplay between agency and structure. Coleman (1988) reminded us that "social capital is productive ... [it] inheres in the structure of relations between actors and among actors. It is not lodged either in the actors themselves or in physical implements of production" (p. 98). Digital capital is therefore a mutual resource so that "features of social organizations such as trust, norms, and networks" coordinate and guide actions (Putnam, 1993, p. 167). Digital ecologies (tools and ways of living in the digital age) both facilitate and constrain, are both actual and potential, and allow for an agency that is bounded within structures in new and compelling ways.

Furlong (2011) unpacked these structures with concepts of connectivity and networks to suggest how digital capital is "part of the process of social reproduction in late modernity." Building on individual practices (e.g., access, skill, competence), structural relations (e.g., connectivity), and the networked society of Castells (2000), he asserted that:

> connectivity is about networks ... modern social structures are organized around electronically processed information networks. Power comes from being networked, and those who are not effectively embedded in networks are at risk of exclusion. Here the favourable contextual conditions that facilitate effective embedding in networks can include opportunity structures, social class, economic resources, etc. In other words, they are the same sort of conditions that are linked to processes of social reproduction in capitalist societies everywhere. (p. 5)

The young people with whom we spoke agreed. If power and status come from being networked, then there are also greater risks of exclusion for those who are not:

WESTON: Yes, I reckon to keep up to date is very important because as soon as you slip or as soon as you start to use out-dated technology, it's difficult then to learn it again.

SCARLETT: I mean like you only have to compare it to people who don't have the technology to see what an impact that it makes in my life ... If you think of Indigenous communities in Australia, it starts like a chain of events you know, like it assists with social education and employment, and there's just so many opportunities that it's great, like it's sort of a necessity at this point and anyone who doesn't have it, like it's such a detriment to just the value of life, I guess, that it makes for a very unequal situation. (Weston, 16 and Scarlett, 16)

If, as Furlong and others suggest, digital capital is a potential feature of complex networked relationships within networked systems, we must acknowledge the array of social and economic influences across dynamic digital ecologies and between humans and machines. We must also recognize the interplay(s) wherein youth animate these ecologies with the tools and networks to which they have access. With this framework in place, we present additional insights from youth participants before offering a model that merges their experiences with these conceptual ideas about digital capital.

Digital capital by/for youth

Young lives provide unique windows into the digital age (Comacchio, 2006; Tilleczek & Srigley, 2017) as they are immersed in exchanges and relations across vast digital networks. Chapter 2 details the methodological and ethical process from which we present here the varied perspectives on digital capital from a sample of young people in Canada, Scotland, and Australia. These young participants utilized digital media to build their social, symbolic, cultural, and monetary capital by enacting their knowledge, competence, and skill. Most common was the way in which they used digital media to find, explore, and apply for jobs towards monetary gain. Nathan (19) discussed how he:

> found a lot of work through the Internet. Sites like Kijiji and stuff, actually I found two good 6–7 month jobs there, like cash jobs, but I mean it was good money and the people were good to me when I did the work.

Mia (18) noted that "you see things advertised on technology like Facebook or something like that and then you can see how that goes," while Sawyer (17) discussed how "from a job market point of view" he "wouldn't shy away from going for an Internet job actually, something that wouldn't entirely require a desk or a place or a uniform." Weston (16) wanted a future job in medical physics and in this context believed that "you have a lot more opportunities with the Internet than without it, so I suppose in that regard that it does have a good impact and improves your life." Camila (17) noted how using digital media provided access to "more information about various careers that you may not have known even existed." Velma (17) provided a concrete example of this when discussing her experience finding out about a camp job through a friend on Facebook, later applying and having all the formalities take place online:

> My friend done it [camp] and he posted photos of it on Facebook. And I never knew that he had went, because I hadn't spoken to him in a while. Then we started talking about it and then that got me wanting to do it. And it was all done online, like, all the application. The phone interview was on Skype ... So I think I would never have discovered that if it wasn't for like, [my friend] putting it on Facebook

Another way that some young people utilized digital capital was by creating potential for commerce. One way this was done was through savvy buying and selling of goods, as exemplified by Sawyer (17). For him, the interplay of digital capital took the form of knowing how to communicate and advertise online, which he said could also be helpful for a future job in retail:

> I remember one of my favourite sales was I paid about £200 for a Nintendo Wii five years ago before I sold that, and these little games I got quite second-hand and cheap. And five years after that I sold it for £220. And that was only because I knew how to sell, I knew how to communicate and I knew how to advertise very well, especially with the basics that I've been used to at art school ... You would do fantastic in

that, a retail selling kind of job if you can do something like that up over the Internet. But I sold plenty at very high prices, but I was very good at selling.

Others recognized different ways to create financial gain, such as Camila (17), who had "heard of ways that you can do surveys and gain money" on the Internet. Natalie (16) thought there was financial potential in playing video games or posting about your life online:

> You can earn money by just playing video games now … people can play video games for a living and they'll get paid by YouTube to post their videos and stuff like that. And there's … Jenna Marvos [You-Tube personality], her life is just posted on YouTube and getting paid for it.

Less common but still important was digital capital for the participants who planned to go into creative industries. For these youth, online self-promotion was crucial and made possible through digital media. Exemplifying this again was Sawyer (17), who was in art school and described using a blog to advertise his portfolio with a view to future employment:

> It's something that I've only just really started from starting art school. It's really just to blog about the portfolio … So if I wanted to become a designer one day or even an architect or anything, that it's important that people can see and view my work and understand what makes me tick as a person, what I'm influenced by, so it's something that I think is very important to update and to really be honest about. Because that will be possibly a pathway into employment one day.

He went on to explain exactly how he used the blog towards these ends, demonstrating that he had a clear strategy and was aware of its importance:

> It should be just professional. I try to put most of my artistic skills, i.e., paintings that I've done, any work that I've done creatively, and try to post that up and just put a little bit about it, write about the time of it. And it's a good way to catalogue exactly when things start and when things finish and really to catalogue how effective they were, and if they were ineffective what can I do next time. So, it's really that, but I don't tend to keep it much like a diary of my own personal experiences, although I do try and have a reflective attitude on my work.

Sawyer was particularly interested in pursuing a career in the technology industry and using his design skills for a large company. He predicted that technology would become more and more ubiquitous and wanted to capitalize on this: "It's what I'll be doing hopefully one day, working for some sort of technology firm like Samsung or Apple, making smart fridges or something like that." He described how smartphones and apps created a passion for technology that paved this pathway, connecting notions of digital capital to his future employment in Big Tech

> Was it five years ago, so I'd been about 12 or 13. But the revolution in the fact that you had Internet on it, that you didn't need a phone line or a Wi-Fi or anything, and you could go on YouTube, you could go on Internet, it had a web browser on it, and it was amazing. And I remember getting updates where you had an application store and that was the year after when the iPhone 3G came out, and I got that one and it was just amazing that you had all these applications that you could branch out your phone and it became a really upgradeable thing. And I genuinely do think that it was something that I spent most of my days thinking about and it was something that I thought myself how I'm driving this as a designer … So it really has actually paved a career path.

Social and symbolic capital towards employment and financial gain were also garnered through social networks, new and old. Some participants were clear about their online networks and how this increased their digital capital. For example, not only did Sawyer (17) showcase his work online, but he also had a large online network that created access to information: "I've got quite a lot of friends. Generally, I think it's very good to network that way, that you add more people, the more contacts you have, the more information you gather." Carson (17) similarly noted how digital media

> helped with networking with people. Yeah, it's easier to network with people. It's probably easier to find services or products you couldn't otherwise find, and it's a good way to find out what people need that you can provide them.

Some participants also were aware of the importance of technology to establish and maintain relations with future clients. For example, Ella (16) described her future and noted, "I'll have like, clients and stuff so technology -- I have to kind of have technology to speak to those people." While not referring to her own clients, Mia (18) similarly recognized the potential for digital networks for her harp teacher: "she is able to publicize herself more and get more clients, not necessarily clientele but just put herself out there more through Facebook and social media and stuff like that because more people use that."

In addition to employment and financial opportunities, participants also spoke about utilizing digital capital for educational attainment and prospects. This included using the Internet to apply to university, which

Ezra (age unspecified) said was "very important because I'm applying to universities through my laptop and stuff. It's really, you know, enabling me do it for each university individually." Harper (17) similarly noted the importance of digital media in learning about and registering for university courses:

> I always see girls on their laptops and they're looking at uni courses, and they sort of teach you that, you know, you've got to look for your courses and there's a little number on it or it has the course number and then you've got to fight for it for you know, and then you've got to put in a preference for early entry or whatever.

Harper also noted how one can use digital media to take part in university without actually attending, opening up opportunities not previously available: "You can even do it externally ... External classes and so you can just do it from home. And so that, that's just using technology and you're just doing a course at home."

In addition to university itself, Scarlett (16) talked about how she was using the Internet to apply for scholarships and how this was "made so much easier" by digital media. She was also applying for student exchange programmes, again noting the importance of digital media:

> I'm currently in the process of applying to like numerous exchange groups that I'm hoping to go on like throughout uni, and without them there would be no way to contact it, there'd be no way to facilitate connections overseas and stuff like that.

Beyond formal education, a few participants also discussed using technology to upgrade their skills and certifications. For example, Weston (16) described completing a "sailing instructor's course" with "the majority" of it being online.

As is clear, participants used digital capital to productive ends and were aware that it was important. While perhaps obvious, it should be reiterated that some technological knowledge and skill is necessary to undertake the activities they described. Participants were aware that having such technological knowledge and ability was vital to personal and professional futures. For instance, Sawyer (17) believed that technology was "growing at an exponential rate" and that "it's something that you have to keep up with every single day to properly understand," while Theo (17) thought that "you miss one thing and you'll just miss it ... [you will] be catching up." Translated into economic terms, individual technological knowledge and competence came to the fore when youth such as Cole (15) noted how "everyone's just using it for everything in most jobs ... jobs at McDonalds, you have to do online to check your shifts." Camila (17) similarly discussed how such knowledge was "certainly important for the future, because I know that a lot of jobs will require knowledge about how to use technologies." In their dyadic interview, Harper (17) and Abigail (17) agreed on the necessity of everyday computational skills "to survive":

HARPER: I think it's important for all ages of people to use it, because things like, just simple things like Word and PowerPoint ... There are so many people who don't know how to use a laptop to its full potential. There's so many things you can do, but a lot of people buy these expensive machines and they don't even know how to use it.

ABIGAIL: They don't, so yeah. I think it's good, it would be good to use it, to know how to use it.

HARPER: In the future, yeah they'll have to learn how to use it.

ABIGAIL: To survive then as well, yeah.

While youth participants were clear that individual technological knowledge and ability was key for future success, they were also cognizant of the fact that there is inequality of access to and knowledge about technology. This is perhaps best exemplified by Weston (16) and Scarlett's (16) exchange earlier, in which Scarlett described how technology "starts like a chain of events you know, like it assists with social education and employment" that creates a foundation of inequality for those without access, specifically referring to Indigenous communities. Scarlett eloquently described how she thought such a "chain" started in poor communities, which ultimately leads to a "cycle of poverty":

Essentially the time, like in poor communities if they're working hard throughout the day then you naturally have less time to come to grips with the rates that you're paying and how to get a good deal and also how to make the most of it. And then there's a lot of other things that feed into it too like in the cycle of poverty, like if you have a poor education, then you're less likely to understand technology, you're less likely to gain employment, less likely to live healthily, etc. etc.

The route to technological inequality was often said to be one's socio-economic status, which then determined pathways to future technical knowledge and expertise:

I would say the cost of technology is very expensive and it is a luxury in many ways. And I would say that people who don't understand technology would also be at a loss again and they would struggle to understand how to use or utilize these smartphones or anything, or the Internet or email. (Sawyer, 17)

Well like I said, it's how expensive it is to have new social media devices and stuff, it's just not everyone can afford that stuff ... I think that the more connected you are and maybe like the higher up you are like, in terms of finance like, financial status and stuff. (Damian, 17)

there would be no money for the phone bill or no money for the Internet bill … when I was a lot younger, so it was like grade 4, 5, 6, it wasn't a big issue. If that happened today, I would be like constantly at the [university] library checking my email and stuff. Making sure that I'm not like missing anything. But I probably wouldn't have gotten those opportunities to be [volunteer executive] because they'd be like "are you always available to do stuff?" And I'd have to be like "well no I have to take a bus to the library to be able to have Internet" and they'd be like "uh we're not sure if we want you in that position." (Marley, 17)

Youth participants also explored potential effects of a lack of digital capital if it cannot be gained through access and technical know-how. Camila (17) discussed her experience growing up in a technologically rich household, juxtaposing this with the possibility of those not benefitting from such an upbringing and subsequently not being able to obtain work:

people of, you know, lower socio-economic areas probably don't have access to good Internet or … good quality laptops or phones and I think that probably sets them back a bit … Because if you don't really know how to use technology, like, I'm fortunate that I grew up with my dad who knows a lot about technology and I'm able to use it quite reasonably but, I can understand people who haven't grown up with that may not be as adept with technology … like for me it's been quite positive but it can impact people quite negatively, especially if they get frustrated and don't know how to use it and feel like there are barriers in getting jobs and talking to people because they just don't, they haven't grown up in, with technology and haven't grown up with people who know how to use it and that can be a problem.

Others such as Ava (16) discussed technological disadvantage in school, which could lay a foundation for future economic difficulty. She noted how Apple computers are expensive and there are "people who obviously financially can't afford it." Yet, at the same time, "the predominant amount of people have that, and our teachers all have it," which meant those who did not have a Mac were disadvantaged because "if the teacher makes a document on Pages or something, then that's not compatible with those [non-Apple] computer's messages to each other." Scarlett (16) similarly said, "I don't think there's enough emphasis about getting equal access to technology, well both in schools and even with getting people to bring your own devices," noting how "in a couple more years, the Year Sevens that come in are going to represent a myriad of demographics and so some people are going to have the latest gadgets and other people aren't." In a particularly salient moment, Damian (17) described the effect that lack of technological access had on people he knew, who eventually dropped out of school:

> I think school should be a bit more sensitive to people in terms of like, in regard to school work and their marks, what they're giving based on how they're supposed to do their work because like, I know some people who, they grew up ... without Internet access at home so it was really hard for them to do anything and over time they just sort of dropped out of school.

Beyond inequality and disadvantage, a few participants also brought forward an implicit juxtaposition between the potential affordances given by digital media and the contradictory practices of large technology companies. This provides one important instance of their macro-political analysis of the digital age. For example, while the use of digital media could vastly improve one's digital capital, Velma (17) was troubled by the fact that technology companies knew so much about users, raising her own Facebook profile as an example:

> And I would probably just go in my profile, because everything is on it, which seems scary when I say it but when you go on "About" it tells you everything, where you're working, who your friends are and where you study, who you're in a relationship with, where you live ... And even though it is still on this Facebook page and anyone, someone that runs Facebook could go see that, really easily, because they can have access to it. That, kind of, scares me a bit.

Carson (17) similarly addressed the macro-political issue of lack of online privacy and felt that it needed to be dealt with before the problem became too big to address:

> as technology starts playing a bigger part in everyone's life, I think the privacy problems are not going to go away but just going to slowly become bigger and bigger until it comes a point where it can't really deal with them unless we sort it out now sort of thing.

Similar notions are found in other participants' statements about privacy and inability to escape from technology:

> It's kinda like I'm imprinted into a system. And it's kind of strange to think about. That you can just type someone's name in and you'll be able to find them online ... it's weird. Well, it's just, it's almost like a privacy issue, you know? (Madelyn, 16)

Big companies are pushing for this point where everyone needs to be connected. Everyone needs to be connected. You got Twitter, you got to tweet you know? You got a phone it's 'gotta say where you are every time you tweet you know? There's no hiding from it, that's what they want. (Carlos, 20)

Also related to privacy were the potential and lasting effects of social media presence on future employment. Some participants noted that both technology companies and potential employers have access to their social media and digital footprints. These young people made clear that one's online presence had the potential to negatively affect job prospects. For example, Leah (16) discussed how "when people apply for certain jobs or whatever like the employer will like search them on Facebook or whatever, and like we were told a story ... about somebody applied for a job and he looked on, like the employer looked on his Facebook page and he didn't get the job because of it." Camila (17) similarly said, "I know that also, potentially [sic] employers have been known to go into people's Facebook accounts and, you know, suss out what they're like."

Towards a model of digital capital by/for youth

Our emerging model (Figure 6.1) makes visible the individual practices (access, skill, competence, knowledge) that young people described as potentially affording them digital capital (symbolic, social, cultural, and monetary) within the structural relations (connectivity, tools, networks) of the digital age. This model is elaborated here as a means of knitting together the lessons learned from youth participants and advancing Furlong's ambition to:

spend some time reflecting on the ways in which the digital revolution can impact on inequalities in advanced societies. Access to digital technologies, competence and skill in using various technologies, and the ways in which involvement in digital networks lead to forms of social connectivity or, conversely, to social exclusion all have the potential to entrench existing inequalities as well as to create new divisions. (Furlong, 2011, p. 1)

Indeed, young people and scholars agree upon the conundrum of digital capital for/by youth. While the potential for access and capital exists, the forms and sustainability of benefits are unclear. Access and skill can be unequal, as these young people observe. They also observe that the lofty ambitions of Big Tech companies run counter to the needs of privacy and well-being (see Chapters 3 and 4 for further explication of these concerns). Thus, the structures of technology undermine the digital capital that youth attempt to develop. These young people recognize that digital tools and

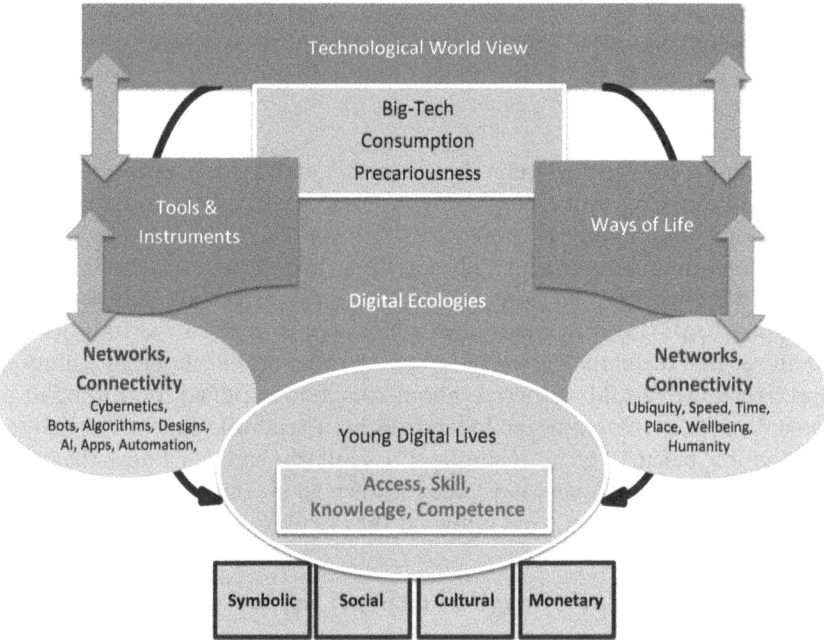

Figure 6.1 Technological World View.

designs are both mandatory and nifty. They also simultaneously question the Big Tech purveyors, their ambitions, and the potential for inequality.

Our model offers a point of entry for analysis of this conundrum. Beginning at the bottom, it illustrates how young digital lives consist of individual and collective *potential* for access, skill, knowledge, and competence that youth use to develop digital capital. Moving from Furlong and others, we posit four aspects that should be analyzed for clearer understanding of digital capital by/for youth: symbolic (ideas, meanings, status, fame, educative experience), social (access, connectivity, people/places in networks), cultural (networks, navigational skill, competence, technique), and monetary (money, employment, upgrading, certification).

Attending to the epistemological fallacy of modernity outlined by Furlong and Cartmel (2007), which suggests that analyses focus on individual action and identity to the detriment of larger structural and social factors, our model expands beyond individual competence and skills. As indicated in the middle of the figure, digital ecologies are comprised of structures and tools (e.g., designs, cybernetics, bots, algorithms, AI, apps, automation) and ways of living in the digital age (e.g., ubiquity, speed, time, place, well-being, humanity) brought by networked connectivity and offered through the technological designs and ambitions of corporations. These ambitions include those perceived by young participants and detailed in this chapter (and throughout this book) as pressures for greater consumption and a loss of privacy and control. Increasingly, commentators and scholars also reflect the concerns of youth through, for example, Facebook's ambitions to build a monolithic but cohesive online community for the "good of users" that runs counter to the well-known breaches of users' data and in favour of the real customers (the advertisers), who, in turn, provide massive profits (Lanchester, 2017; Vaidhyanathan, 2018).

The youth in this chapter are brokering digital tools and networks for employment as they must. However, they do not speak to the force of precarious labour for young people that is being brought by automation (Carr, 2014) and a host of economic, political, and social factors in late modernity (MacDonald, 2017). Nor do they yet mention the global growth of youth underemployment and precarity in the digital age even though "scholars have interpreted the growth of precarious work as indicative of an important social change: the emergence of a new social generation or the rise of the new Precariat class" (MacDonald, 2017, p. 162). Thus, precariousness must be considered alongside Big Tech capital and consumption as features of the future analysis of digital capital by/for youth, as is indicated towards the top of our model.

This is the technological worldview which young people navigate and is illustrated at the top of the model. This worldview has been detailed by scholars (Tilleczek & Srigley, 2011, 2017) as that which brings together the vampirism of modern capitalism with the ambitions of continuous improvement, perfection, and amelioration that leads the technological revolution. As Lanier (2010) describes the view, machines do not *make* human life better; they *are* better than human life. This view of life has far-reaching economic, political, and social consequences, not least of which is disappearing and precarious labour (Carr, 2014). When aimed at youth, this worldview delivers ecologies (tools, devices, ways of life, and self-images) as fundamental, emerging contexts within which they attempt to secure a place in work, education, and society. The feedback-loops in our model are signified by arrows to show the ongoing interplay(s) as productive spaces of agency that young people animate to negotiate and manage these ecologies. The interplay(s) are openings for young citizens to develop and design their own tools, spaces, knowledge, and networks that have the potential to afford them digital capital. Furthermore, they are openings into encounters with new forms of inequality and precariousness that they negotiate with their digital capital.

Conclusion

> **Mark Zuckerberg**: The thing about the ad model that is really important that aligns with our mission is that — our mission is to build a community for everyone in the world and to bring the world closer together.

And a really important part of that is making a service that people can afford. A lot of the people, once you get past the first billion people, can't afford to pay a lot. Therefore, having it be free and have a business model that is ad-supported ends up being really important and aligned. Now, over time, might there be ways for people who can afford it to pay a different way? That's certainly something we've thought about over time. But I don't think the ad model is going to go away, because I think fundamentally, it's important to have a service like this that everyone in the world can use, and the only way to do that is to have it be very cheap or free. (Roose & Frenkel, 2018)

Some of the youth participants in our research knew well what Zuckerberg was suggesting and questioned it, evident in their statements about the problematic practices of technology companies. This echoes concerns raised by Furlong and others who are asking: how are young people actively negotiating (collectively or individually) the ecologies that make up the digital age? What are they up against and what does this mean for inequality and agency? The digital age continues its strong march forward into the lives of young people, and we have witnessed their attempts to use an array of networks and tools to their benefit.

However, the young people with whom we have spoken detail a conundrum. They are both hopeful and critical. They participate in digital ecologies while attempting also to see ahead to what the digital age has in store for them and how they might harness it. Yet, most days are dizzying when it comes to keeping up with these trends. Our model begins to set out a path for analysis, asking that we notice where technology has been, where it is headed, and how young people attempt to capitalize upon it while it capitalizes upon them. We concur with and build upon Furlong's analysis of digital capital in late modernity:

> Tied into the equalities agenda on a broader political level, there is also potential to develop a greater understanding of the ways digital technologies promote new forms of political engagement involving groups that were once denied a voice and allowing citizens to challenge power in new ways. (Furlong, 2011, p. 7)

However, our model is a visual reminder of the possible ways and extent to which youth are being denied power and opportunity, of the new forms of inequality and social reproduction in young lives, and of the productive work that youth are undertaking to navigate it all:

> The technology is increasing, obviously, but the dynamic of people who have lesser technology and greater technology, that's not changing. That's not changing. Like, people are getting better cell phones and

whatnot, but then there's all these people who are getting even better cell phones. So nothing is actually changing. (Jace, 17)

It's not like when they enter the Internet it's, like, everybody is equal on the Internet. It's like you're still sitting on your computer and whether you're poor as fuck and living in some shitty trailer or something maybe in the middle of nowhere, but you've got Internet access. It's now, like: hey, I'm up there with the haves, but you're not. I mean, it sounds really rude, but it's the truth, I guess. (Carter, 18)

Notes

1 On February 14, 2018, a former student entered a high school in Parkland, Florida, USA and killed 17 people (https://www.cnn.com/2018/02/18/us/parkland-florida-school-shooting-accounts/index.html).
2 Facebook, Google, Apple, Amazon, and Microsoft (Lotz, 2018).

References

Bourdieu, P. (1986). The forms of capital. In J. P. Richardson (Ed.), *Handbook of theory and research for the sociology of education* (pp. 241–258). New York, NY: Greenwood Press.

Carr, N. (2014). *The glass cage: How our computers are changing us*. New York, NY: WW Norton.

Castells, M. (2000). *The rise of the network society* (2nd ed.). Malden, MA: Blackwell.

Coleman, J. S. (1988). Social capital in the creation of human capital. *American Journal of Sociology, 94*(Supplement), S95–S120.

Comacchio, C. R. (2006). *The dominion of youth: Adolescence and the making of modern Canada, 1920 to 1950*. Waterloo, ON: Wilfred Laurier University Press.

Furlong, A. (2011, Dec). *Digital capital and inequality in late modernity*. Paper presented at the Future of Youth Sociology Symposium at the Australian Sociological Association Conference, Sydney, Australia.

Furlong, A. (2012). *Youth studies: An introduction*. London, UK: Routledge.

Furlong, A., & Cartmel, F. (2007). *Young people and social change: New perspectives*. Buckingham, UK: Open University Press.

Lanchester, J. (2017). You are the product. *London Review of Books, 39*(16), 3–10. Retrieved from https://www.lrb.co.uk/v39/n16/john-lanchester/you-are-the-product

Lanier, J. (2010). *You are not a gadget: A manifesto*. New York, NY: Alfred Knopf.

Livingstone, S., Lemish, D., Lim, S. S., Bulger, M., Cabello, P., Claro, M., ... Nayar, U. S. (2017). Global perspectives on children's digital opportunities: An emerging research and policy agenda. *Pediatrics, 140*(Supplement 2), S137–S141.

Livingstone, S., Mascheroni, G., & Staksrud, E. (2018). European research on children's Internet use: Assessing the past and anticipating the future. *New Media and Society, 20*(3), 1103–1122.

Lotz, A. (2018, Mar 23). 'Big Tech' isn't one big monopoly – it's 5 companies all in different businesses. *The Conversation*. Retrieved from https://theconversation.com/big-tech-isnt-one-big-monopoly-its-5-companies-all-in-different-businesses-92791

MacDonald, R. (2017). Precarious work: The growing précarité of youth. In A. Furlong (Ed.), *Routledge handbook of youth and young adulthood* (2nd ed., pp. 156–163). Abingdon, UK: Routledge.

Organisation for Economic Cooperation and Development [OECD]. (2011). *Against the odds: Disadvantaged students who succeed in school.* OECD Publishing. doi:10.1787/9789264090873-en

Putnam, R. D. (with Leonardi, R., & Nanetti, R. Y.). (1993). *Making democracy work: Civic traditions in modern Italy.* Princeton, NJ: Princeton University Press.

Riley, N. S. (2018, Feb 11). America's real digital divide. *New York Times.* Retrieved from https://nyti.ms/2BUVw9a

Roose, K., & Frenkel, S. (2018, Mar 21). Mark Zuckerberg's reckoning: 'This is a major trust issue'. *New York Times.* Retrieved from https://www.nytimes.com/2018/03/21/technology/mark-zuckerberg-q-and-a.html

Statistics Canada. (2018). *A portrait of Canadian youth.* (11–631-X). Retrieved from https://www150.statcan.gc.ca/n1/pub/11-631-x/11-631-x2018001-eng.htm

Tilleczek, K. (2011). *Approaching youth studies: Being, becoming, and belonging.* Toronto, ON: Oxford University Press.

Tilleczek, K. (2014). Theorizing young lives: Biography, society, and time. In A. Ibrahim & S. R. Steinberg (Eds.), *Critical youth studies reader* (pp. 15–25). New York, NY: Peter Lang Press.

Tilleczek, K., & Srigley, R. (2011). Modern youth at work and play. In K. Tilleczek, *Approaching youth studies: Being, becoming and belonging* (pp. 66–86). Toronto, ON: Oxford University Press.

Tilleczek, K., & Srigley, R. (2017). Young cyborgs? Youth in the digital age. In A. Furlong (Ed.), *Routledge handbook of youth and young adulthood* (2nd ed., pp. 273–284). New York, NY: Routledge.

UNICEF. (2016). *The state of the world's children 2016: A fair chance for every child.* New York, NY: United Nations Children's Fund. Retrieved from https://www.unicef.org/publications/files/UNICEF_SOWC_2016.pdf

Vaidhyanathan, S. (2018). *Antisocial media: How Facebook disconnects us and under-mines democracy.* Chelsea, UK: Oxford University Press.

Wilkinson, R., & Pickett, K. (2009). *The spirit level: Why equality is better for everyone.* New York, NY: Penguin.

Chapter 7

Digital media, youth, and social relationships

Jonah R. Rimer and Kate C. Tilleczek

Introduction

> I think we should be taught how to use it at schools. Here's the Internet
> and it's a whole different way to talk to your friends. (Abigail,[1] 17)[2]

Digital media has brought significant shifts in social interaction. Contrary
to utopian or dystopian visions, such shifts are complicated and bring
changes and challenges that can be interpreted as simultaneously positive
and negative. Having lived online more than any other generation, our
youth participants are at the forefront of these changes; as noted by Sawyer
(17), "to experience the modern world you would have to have Internet cause
that's what the modern world is." Focussing on participants from Scotland
and Australia (43 young people in total), this chapter centres on young
people's experiences and perceptions of online social landscapes, norms,
and interactions.

Participants describe a plethora of new advantages, horizons, and
possibilities brought to them by digital and social media: convenience, availability, and speed; social organization; better forms of communication;
transcending geography; belonging and support; novel, ongoing, or rekindled connections; and enhanced relationships and social lives. However, they
simultaneously discuss a multitude of concerns and difficulties: anonymity
and detachment; exclusion; representation, competition, and striving for
attention; pressures and anxieties; visibility, scrutiny, and judgement; difficulty with relationships and social skills; misinterpretation and meaning; less
face-to-face interaction; and suffering relationships and social lives.

We unpack these themes by organizing youth perspectives into three
sets of tensions: (a) *easy reach vs difficult distance*, (b) *forging connections
vs changing relationships*, and (c) *enhanced vs suffering sociality*. It is then
apparent that digital media use is bringing with it many shifts to the social
lives of young people that participants experience and perceive as positive
and negative, helpful and hindering, and valuable and detrimental: as

exemplified again by Sawyer (17), "that's just the contentious part of most of the things that evolve with life, that it will have good and bad points." Social relationships for youth in the digital age are no exception.

Theoretical framework

Before presenting youth perspectives, a short theoretical summary is necessary to facilitate our analysis of social relationships in the digital age. We follow the call to anthropologists to analyze "the cultural constructions and reconstructions on which the new technologies are based and which they in turn help to shape" (Escobar, 1994, p. 211). Anthropologists put forth two relevant positions about online technology mediated relationships and interactions. First, activities, behaviours, and social constructions on the Internet relate to and affect people in the offline world, and these contexts are not dichotomous or entirely discrete (Axel, 2006; Barendregt, 2013; Kirmayer, Raikhel, & Rahimi, 2013; Miller & Slater, 2000; Wilson & Peterson, 2002).

Second and simultaneously, online spaces are distinct, different, and constructed and interpreted as such by users (Miller & Horst, 2013; Miller & Slater, 2000). Scholars suggest that social norms in given parts of the online world can be related to, similar to, or divergent from other online spaces, as well as related to, similar to, or different from offline spaces (Budka & Kremser, 2004; Slater, 2002). The idea that digital is opposite to real is rejected (Boellstorff, 2013, 2016); instead, virtual spaces are "real places that must be understood in their own terms" (Boellstorff, 2016, p. 395). In short, online spaces and those inhabiting them have and produce unique interactions, norms, and relationships.

With these notions in place, the remainder of this chapter focusses on addressing two questions: (a) how do young people's online social lives meaningfully connect to their offline lives, states of mind, and behaviours, and (b) how do young people experience norms and forms of sociality in online spaces? We posit that the answer to both lies in the negotiated tensions between and among online and offline socialities and social relationships.

Youth perspectives

Easy reach vs difficult distance

For the youth participants in our study, the first set of social benefits brought about by digital media centred around convenience. Most common was a reported ease of communication and availability: "Just in being able to, like, talk to anyone really, whenever and wherever you are" (Cole, 15). Many participants focussed on how digital and social media afforded better and wider communication: "And I always say like with social media that people can get in contact a lot easier. Personally I wouldn't imagine, I can't even remember how I used to keep in contact before Facebook" (Sawyer, 17). In 36 interviews, this idea was elaborated as the ability to keep in touch, often with people living elsewhere. For example, Harper (17) noted:

> It's good just to be able to interact with people, a couple of people who went to school and then they were really close and they live in [another city] now so I can talk to them and see what they're up to ... Just stay in touch because that's how you keep friendships.

Some also found that digital media helped them stay in touch with people they saw day-to-day but with whom they had missed a social occasion: "if you can't go somewhere you can keep up with people really" (Audrey, age unspecified).

As Harper pointed out, keeping in touch often involved transcending place. This was a common theme in 25 interviews, where participants focussed on how communication across geography was made possible through digital media. For example, Ella (16) noted that "my uncle who lives in [another country], I can, cause we're, we're pretty close ... contact him whenever I need to," while Camila (17) talked about how "I've got relatives in [another country] and all over the world, it's easy to connect with them." Three participants in particular described how digital media "made the world smaller" (Carson, 17). Also related was speed when engaging socially. For example, when transcending geography, Camila (17) noted, "[it] helps stay connected with people around the world ... in an immediate sense, not, like, a lag, no letter writing or anything like that these days."

Another common theme related to convenience found in 24 interviews was the assistance that digital media provided when scheduling and making plans to socialize. For example, Cole (15) described how his main use of technology was "to make plans" which included surfing and skateboarding with friends, while Abigail (17) noted how she "never related a lot of my fun times with technology" but that it was "more about organizing it on technology or sharing after."

However, tensions became apparent when participants elaborated on problems they faced. As discussed, anthropological theory posits that online activities and experiences relate to and have effects on offline activities and experiences. Actions on the Internet can become important parts of everyday life (Miller & Slater, 2000), and use of the Internet has potential to affect and be affected by offline relationships (Kirmayer et al., 2013). Indeed, young people described how conveniences brought about by online communication also had consequences for their offline lives, experiences, and emotions: there was seepage in and across these ecologies.

While digital and social media afforded the ability to keep in touch, communicate, and be accessible, some participants also felt the need to always be with technology; the ability to always be in touch also meant that some struggled when digital media was not available. Fifteen interviews had participants speak about anxiety when not with their phones, 12 included the notion of having a *digital leash* tugging at one's neck, and eight included mention of the fact that participants could not imagine life without phones. Ezra (age unspecified) said that if he had to be without a phone for a week, his dad "has a work phone ... so I'd take the extra one." Velma (17) discussed how "even if I don't have my phone or I lose my phone I, like, panic," while Abigail (17) described her state of mind when her phone runs out of battery as, "You just feel so helpless, like what the hell am I supposed to do now?" In her interview, Ariana (age unspecified) came to the realization that her phone was with her at all times, noting,

> my phone's like the only thing that I will keep with me ... actually I didn't realize how much I have it ... If I get home from school I'll sit at my desk and then I want to go upstairs, I'll probably take it with me.

Audrey (age unspecified) described how she "misses" her phone, "needs to go on it," and would not go as far as her garden without it, also asking if the interviewer "felt lonely" when without a phone. Most indicative of this anxiety, during her interview Luna (17) tried to connect to the Internet. She felt unsettled when disconnected and described a loss of Internet as "torture and you have no data on it. And you're just sitting there ... I just feel like my world is ending."

Pressure can also come with better and faster modes of connection, and eight participants made clear how constant availability and connectedness were stressors. This included being too accessible, feeling pressure to keep up with updates, believing that online communication was too fast, and feeling pressure to respond to messages and engage even if one did not want to. For example, in their dyadic interview, Abigail (17) said she "hates how people can get to you all time," to which Harper (17) responded by recalling her frustration with a friend: "if she's bored she'll call me and want to chat and like sometimes I just like I want to be home by myself." Damian (17) contrasted how "it's really good to be able to like, have things here and now" with the fact that "now it's like if you don't see that update or something in the next few minutes like, you're a bit late, you're a bit behind." Camila (17) was most blunt about how being constantly connected and available some-times affected her relationships, notably using the word "escape":

> sometimes you feel pressured to talk to people because they're there and you're there and sometimes it can be hard to escape, unless you log-off, but then sometimes that's not an option because you're talking to one person and sometimes you, OK if I'm gonna be honest there are some people you don't really want to talk to all the time and then that becomes a problem.

The ability to transcend physical space (geographical reach) was seen as another benefit to digital communication. Yet, tension was again apparent in that *geographical* reach was positive, while *digital* reach was not always so (even though digital reach provides the possibility for geographical reach). For instance, some participants noted how digital reach was intrusive such that arguments could be broadcasted online and everyone could see comments, meaning that personal opinions and issues no longer stayed private which "makes everybody aware of it rather than keeping it between however many people it has to be between" (Leah, 16). This negatively impacted social situations: "So people were asking and like everyone found out when it should have just stayed private. And that happens a lot ... there's arguments and because everyone can see it, more people get involved and take sides" (Ariana, age unspecified).

Similarly, 12 interviews had participants speaking about being too visible such that many people can become aware of personal issues, and thus privy to past mistakes. For such reasons, Scarlett (16) thought it was "important to present yourself online in a way that you wouldn't mind if say your grandmother saw you on there." Best exemplifying this is Xavier (16), who noted that "lots of people can see you so that there is no way of getting out," elaborating:

> But I went out with a girl, doing something inappropriate, and that went all over the school ... and I think all over [the country] and even further just from the Internet. And then there was another time where I was in an argument with someone, and like the school, everybody got to see it, and it becomes a thing that you are stapled with. Like, I'll be known as the person that argued with such and such about what topic.

Finally, while digital media was seen as helpful for scheduling and making plans, tensions appeared when participants described exclusion. Some said that social media was a necessity and people were excluded without it: "I don't think there's really anyone except for one person that I know that doesn't have a phone ... And the only time I meet them is if I ever see them out by chance" (Aurora, 19). As social media is ubiquitous, this effectively limited participants' choice to engage with digital media or not. Exemplifying this is Scarlett (16), who noted that "technically there's a choice, there's always a choice. Realistically, there isn't really." She said that choosing not to use social media could be socially "disastrous":

> if I shut myself off from Facebook and didn't go on it for I don't know like a year ... unless I was prepared to talk to people and organize things and meet up with them, things would happen like, events organized online, inevitably I will be forgotten when those events are being planned you know? Because like, one day someone will forget to

text me so I won't go and then I won't be invited to the next and then the next thing so it's sort of dominoes like, you have a choice but you're definitely sort of nudged towards one of them.

Forging connections vs changing relationships

A second set of themes that emerged as new advantages and possibilities revolved around developing and continuing relationships. Some participants noted how digital media made it possible to foster friendships with people met offline. This could be through travel, such as with Imelda (17), who remarked that she "went to [another country] last year, and keeping in contact with our host sister has been pretty great over Facebook," as well as with Velma (17), who said that "if I meet people on holiday, I'll usually come home and I'll still talk to them, like, on Facebook or something." This facilitation also occurred with day-to-day activities, such as with Luna (17): "we met at my cousin's party. But we kept in contact with that and then we just came really, really, really close friends on Facebook. And that's always been there, like, to keep us as friends."

Another way that digital media facilitated relationships for three participants in particular was reconnecting with friends and family. Camila (17) discussed this in her interview:

after the [country] Revolution, most of my family kind of split up and spread. And so some of them moved to [another country]; we've only just recently, very recently, connected with them and that was partly because of Facebook and we found them through that.

Ava (16) similarly noted how Facebook helped her reconnect with childhood friends, also invoking geographical reach:

Well, in terms of being overseas in [another country], I still have a lot of contact with a lot of my friends. Because we were just writing letters to each other, and then we sort of dropped out of that, because it's a bit of a hassle to do that all the time. And then I found them again through Facebook, that's nice.

In terms of creating and maintaining relationships not involving the offline world, 15 young people discussed how digital media was used to make new friends, as well as to meet and communicate with new people. Sometimes this was mutual friends of friends, while at other times not. Ella (16) discussed how "you can connect to all different people on it like, it's kind of interesting like, and you, meet new people on it." Again invoking geography, Imelda (17) talked about how digital media "can open you up to like a whole other world and allow you to connect with people from all over the world," while Xavier (16) elaborated on how he became "best friends" with someone he met online: "I have a friend in [another country] I have never met before, but she's nice and we've talked for almost a year now. But I have never met her in real life but it seems like we're best friends."

As Boellstorff (2013, 2016) notes, relationships, norms, and features of interaction in online spaces can be distinct from other contexts and should be treated as such. Through this lens, while making and continuing relationships was a new possibility through digital media, tensions appeared when participants discussed the changing nature of social relationships and interactions. In 18 interviews, youth talked about no longer spending time outdoors or being active with friends; in essence, sociality moved indoors. The most striking example came from Xavier (16) who was asked, "If your life was a book, what would you name the main chapters that include technology?" He responded: "Time I should have just spent elsewhere." Scarlett (16) similarly spoke about how she now "had to deal with trying to coerce people out of their homes or answer the phone to just talk to me." Others, such as Velma (17), contrasted present-day relationships to childhood:

> When I was younger it was more, like, you were outside and that was that. There was no, like, going on Facebook and, or tweeting or whatever, you were outside with people … And then when I started high school it was when all the technology kicked in.

For participants, digital media facilitated and fostered social connections, but simultaneously created another tension through the experiences of such relationships. For instance, 21 interviews included youth speaking about how digital media use resulted in less face-to-face interaction. In fact, 17 reported not having authentic relationships. Damian (17) spoke to this when he said that digital media "stops people from, kind of, socializing a bit more like, it's supposed [to be] social media but it doesn't really help that much … in real life situations." Scarlett (16) believed that "you can't really have a deep and meaningful conversation online," while Camila (17) suggested that her social skills were changing:

> Sometimes I feel because I spend so much time online that talking face-to-face probably doesn't come as easily as it might if I didn't have the Internet because I'm so used to talking and expressing myself through, like, writing rather than verbally.

Finally, Abigail (17) provided an example when recalling an awkward car ride:

> I was dropping someone off once and they just didn't really want to talk and they were just scrolling on their phone. And it was just like an awkward silence and I was just like, "Ah, why?" No, it's just rude. It's

a little bit rude I think. If they just prefer to do whatever they want to do instead of actually talking to someone in conversation and having quality time, that yeah, I always think it's a bit too much.

The ideas discussed by young people help us begin to unpack how alterations in social interactions and norms are experienced in the digital age. With this in place, we turn to deeper understandings about sociality and the ways in which it is experienced by young people as both enhanced and retracted.

Enhanced vs suffering sociality

Participants described many ways in which digital media improved their social lives. In 16 interviews there were detailed discussions about social media enhancing sociality, and in 15 interviews youth participants said that digital media made them more social and interactive. Perhaps the best example was Weston (16), who articulated how social media enhanced his social life while also invoking many other themes discussed thus far (e.g., geography, making new friends, planning events):

> I mean, so I had the friendship group at my school of people that I knew and people that I talk to, but I only really knew personally one or two people, and I only used to go out with those one or two people and just because I didn't really go out much and I was just at home doing stuff ... As opposed to when I started using Facebook and when I started ... having conversations with other groups at school and at other schools, I used to start going out more and learning more about people and doing stuff because we'd organize to do something as a group and go to the movies or go to the beach, so I gained a lot of good friendships because of that.

Other youth raised further examples of how digital media had a positive effect on their social lives. In 13 interviews, participants spoke about how digital media helped them to connect and belong, such as Ezra (age unspecified) who described how it

> Broadened my friends as well, got more friends through that ... like people in my school that like the same things as me ... And I found out online that they did and I was like oh, got some common ground.

In seven interviews, participants discussed how current social relationships were enhanced by digital media by allowing for more contact. Scarlett (16) noted how "I wouldn't have the same kind of relationships I do with my friends because I wouldn't talk to them outside of school," while Ella (16) similarly said, "I reckon it makes you closer as friends, cause then you can stay connected [to] each other outside of school"

Another way in which digital media improved social lives was through the ability to stay updated. In 15 interviews, youth talked about digital media making it easier to keep up to date with other people's lives and/or keeping others updated about them. Ezra (age unspecified) shared "wedding photos" and "photos of my nephew" with his family abroad, while Ariana (age unspecified) discussed how digital media created a feeling that one has continually seen people, even if one has not:

> there's someone that's left college and I maybe will see them once a month, or less than that, but I'm still like keeping up to date. It's not like you haven't seen them in two months and then they tell you what happened.

Similarly, four participants talked about how digital media allowed them to share good news. For example, Velma (17) described how "when I got my certificate and for my degree, I put it up on Facebook and everyone was, like, 'Oh, I didn't even know that you done that,' I was like there you go."

Although mentioned by fewer participants, a final way that digital media improved social lives was by facilitating more genuine and supportive relationships. Cole (15) thought that interacting online without being seen made it easier to be honest: "Well it might be easier to talk to someone online ... Or through Facebook or through Skype without them actually seeing you ... Just easier to I guess tell them what you want to." In five interviews, participants discussed interaction on digital media allowing them to support or be supported by others. Ezra (age unspecified) talked about how "if someone's having a hard time and they post about it on Facebook, you can show your support for them," while Camila (17) noted, "I'm able to talk to my friends, if they have a problem or if I have a problem we're always there to support one another because we're mostly online at the same time." Finally, Scarlett (16) discussed how providing support had a positive effect on her relationships:

> I mean there's been times where ... my friend's needed a lot of help and so because Facebook was a thing she could just message me and so we could organize to meet up instantly and I could help her out and vice versa, and so there's been a lot of positive experiences with stuff like that like if you know you're right on the edge and you really just need someone to talk to they can be there for you.

However, tensions became apparent when participants discussed online norms and interactions simultaneously causing a multitude of problems. While some young people said that digital media provided enhanced sociality, more of a social life, and a sense of belonging and connection, some also discussed how their sociality and social lives were suffering. In nine interviews, young people talked about their close relationships failing or being negatively affected by technology use. This is best exemplified by an exchange between Camila (17) and Imelda (17), in which they described spending less time with family:

> IMELDA: Spending less time with family at home ... So I'll be either in my room, or even if I'm out in the dining room, I'll be pretty much on my laptop a lot of the time.
>
> CAMILA: Yeah, sometimes it can be pretty frustrating because my dad really enjoys technology, he's on his phone quite a bit, so I'll try to talk to him and he'll just be like sitting there looking at the news.

Specifically tied to face-to-face interaction, participants in six interviews suggested that offline communication had become more difficult after becoming accustomed to online interaction: "Just because you're more, like it's more easy to talk to people online than what it would be face-to-face. You can't hold a conversation, so stuff like that really" (Audrey, age unspecified). Similarly, in six interviews participants brought up the idea that due to so much interaction happening online, they or others were lacking social skills. As noted by Scarlett (16):

> I mean if you don't use it in the right way it can potentially be rather a destructive force like, if you, if you're going to have trouble conversing with people and then you come to rely on technology you're going to find it difficult throughout life because you'll always have to do things face-to-face and if you keep taking the easy way out then it's going to backfire on you eventually.

Aligned with anthropological positions, youth also described a multitude of unique norms and practices associated with online sociality that they perceived as problematic. While some experienced relationships mediated by digital media as more genuine and supportive, there were many discussions of the opposite experience. The first of these centred around the idea of a *digital cloak* and the ability to hide behind it. Key to this was anonymity and meanness that can arise from it.

In the early to mid-2000s, most online interaction was text-based (Barak, 2007; Gackenbach & von Stackelberg, 2007), while current research suggests an increase in visual communication with the use of applications such as Instagram and Snapchat (Gruzd, Jacobson, Mai, & Dubois, 2018; Smith & Anderson, 2018). At the time of data collection, our participants engaged mostly textually, but also visually, with Facebook usage being most common, followed by texting, email, Instagram, Snapchat, and Reddit. With text-based communication, elements used to identify people in offline settings (e.g., gender, ethnicity, age, height, weight, accent) are not easily visible online when all one can see is text; the same is true of many social cues normally present in face-to-face interaction such as posture, tone of voice, body movement, facial expressions, and eye contact (Amichai-Hamburger, 2007; Brignall III & Van Valey, 2005; Christopherson, 2007; Hardey, 2002). As such, people can construct how they would like to be identified (Ellison & boyd, 2013) or openly display hostile behaviour under the guise of an online character (Brignall III & Van Valey, 2005).

Participants experienced this, or knew of others who had, and discussed the impacts of online anonymity. Carson (17) believed that "anonymity creates problems all the time" and that "it's only ever been used negatively as far as I can tell, in my experience." He thought that anonymity "brings out the worst sides of people because there's no real immediate threat and they can be as cruel as they want to, and that quite often comes out." In 10 interviews, participants also brought up the fact that online, people will say or do things that they would otherwise not. Exemplifying this is Audrey (age unspecified), who thought that "people are too scared to like say stuff in person ... They prefer to do it over the Internet because they feel as though nobody would catch them out." Also speaking to the notion of a digital cloak and text communication, Ella (16) noted that online, one can act differently to how one really feels. She described how "even if I'm talking to a friend I can act so different to how I like, am, or how I'm feeling like that I suppose ... you can do whatever you want, you can act however you want."

In addition, while awkwardness and appropriateness may not allow people to exit a face-to-face conversation whenever they please, online textual interaction instead permits individuals to log-off at any point (Amichai-Hamburger, 2007; Gackenbach & von Stackelberg, 2007). Users can move in and out of interactions freely, which can result in abandoning conversations as opposed to investing time in resolving issues (Brignall III & Van Valey, 2005). Three young people spoke to this, noting how it is easy to disengage:

> I talk to most of my peers on social media at some point or another. Cause it's easy to get in contact with them if I want to, and it's just as easy for them to disappear if they don't want to talk to somebody else. (Carson, 17)

Without confirming body language, online text-based communication can require more interpretation from users (Bargh, 2002). This was another area of difficulty for some young people; contrasted with face-to-face communication, participants in eight interviews described how tone, emotions, and intentions can be misinterpreted online, leading to arguments:

> But then on the reverse side there have been times that someone's said something and if it was in person the tone of their voice would have

made it clear what they meant but online because people are in a bad mood you know you don't have the time to just sit back, reflect on what they meant, think rationally and then come back, you just type the first thing that comes to mind and the result can sometimes be quite messy like, you know a complication that has to be sorted out. (Scarlett, 16)

As described earlier, this was largely due to the fact that textual interaction was missing sensory and social cues normally found in offline communication. An example was provided by Camila (17), who discussed a fight she had with a friend because of the inability to hear intonation:

people can misinterpret things online because you can't hear the intonation in people's voices, you can't, like, hear their expression or whether it's facetious or whatever, and so people can misinterpret things, like, she misinterpreted something and she got really mad and kept sending me some hateful messages and, yeah that got a bit annoying but, then once we talked about it face-to-face it was all sorted out, it was fine.

In a final area of tension, while some youth described digital media helping them stay up to date with others through posts, participants also discussed difficult and problematic norms and practices associated with posting. First was scrutiny; some spoke about how online content is "scrutinized so much" (Xavier, 16), which could make them self-conscious about what is posted: "So she'll tag me in statuses and I'll remove the tag ... like, baby photos where I'm looking chubby and I'm like, 'remove that tag now'" (Luna, 17). Indeed, with widespread use of smartphones, some scholars suggest that people are now carrying a "portable panopticon" (De Saulles & Horner, 2011), with opportunity to continually monitor others. Young people reflected on such monitoring, with five interviews including discussions regarding the need to think carefully about how to articulate oneself in the "unforgiving environment" (Scarlett, 16) of the Internet. For example, Carson (17) noted how he self-censored because of scrutiny from peers: "And it's constraining in that when I say something on Facebook I have to think very carefully about how I'm going to say it ... has to be censored, almost." In one particularly honest moment, Ella (16) encapsulated these notions when she discussed judging people based on their posts:

lots of girls put up like, silly photos like, in their bikini and stuff like, I don't know, I just wouldn't feel comfortable with that because I know anyone can see that ... they can take it and interpret it in any way they'd like it to interpret and people can judge you on that sort of stuff as well like, how you portray yourself like, through images you upload because ... honestly I do judge people like, on how they portray themselves like, on Facebook and stuff like that.

Also related to scrutiny and having posts watched, a second set of norms and practices that some young people found troubling centred on others' attempts at garnering attention. One aspect of this was competition for attention, something on which Xavier (16) focussed. Invoking notions of evolution, he suggested that social media makes people "go back to their primitive nature and like try to be the main person, the alpha dog and try to be the most popular person" He said that "the amount of re-blogs you get, the amount of viewers you get on YouTube, it matters and it creates a lot of competition. And it's really pathetic, it's kind of like ... crying for attention." The idea that people were "crying for attention" via posts was also something Natalie (16) alluded to when she claimed, "I see it just about every day on Instagram; people post their selves naked for likes," while Damian (17) similarly noted how "it makes it a lot easier for people to pretend to be something they're not ... you always see people like, just seeking attention and crying out for help." In a particularly interesting moment, he told a story about younger children "missing a big part of their childhood" by striving for "likes": "I was on the bus the other day and I heard these kids talking about how many 'likes' they've got on Instagram and stuff I'm like, really? You're in primary school like, have fun while you're young."

The idea that people are in a constant online popularity contest was not confined to participants referring to others; some also described their own efforts. For example, Leah (16) talked about her strategy for choosing photos to post and making sure they looked fun and not staged:

> It depends on how bad they might have been ... Like there's a few pictures that I know that friends have got that have just been taken in ... a certain face or whatever, and you kind of sit there and go "do not put that on." But other than that, like not for like self-conscious reasons really ... Quite a lot of the pictures that we take are all just uploaded anyway just because it just makes it look a lot more fun, it doesn't make it look like we're posing all the time kind of thing.

Similarly, Ella (16) described how she used to be "boy crazy" and was happy when good-looking boys "liked" her photos: "I used to be like, heaps like, boy crazy when I was younger [laughs] ... if lots of like, attractive boys 'liked' my photos I'd be, you know, all happy about that but now I don't really [laughs] care." In all this, it can therefore be seen that young people experience and perceive online interactive and social practices as sometimes enhancing and positively contributing to sociality and social lives, while sometimes being a negative force.

Conclusion

Informed by theory from anthropology, this chapter explored tensions in young people's experiences and perceptions of online sociality, social norms, communication, and interaction. The tensions that youth illuminate include *easy reach vs difficult distance, forging connections vs changing relationships,* and *enhanced vs suffering sociality.* In particular, young people experienced and perceived multiple positive elements to digital media, including social organization; better communication; transcending place; belonging and support; novel, ongoing, or rekindled connections; and improved relationships and social lives. Yet they also outlined concerns and difficulties: anonymity, exclusion, striving for attention, stress and anxiety, exposure, scrutiny and judgement, misinterpretation, less face-to-face interaction, and suffering relationships and social lives. These tensions suggest that social interaction and sociality in the digital age is complicated, shifting, and at times contradictory, requiring young people to negotiate novel and uncharted social realities.

Notes

1 All participants have been given a pseudonym.
2 Quotations have been edited for readability: expressions such as "um" have been removed, and ellipses indicate when a word, sentence, or section has been removed but the meaning of the quote has not changed.

References

Amichai-Hamburger, Y. (2007). Personality, individual differences and Internet use. In A. N. Joinson, K. Y. A. McKenna, T. Postmes, & U. D. Reips (Eds.), *The Oxford handbook of Internet psychology* (pp. 187–204). Oxford, UK: Oxford University Press.

Axel, B. K. (2006). Anthropology and the new technologies of communication. *Cultural Anthropology, 21*(3), 354–384.

Barak, A. (2007). Phantom emotions: Psychological determinants of emotional experiences on the Internet. In A. N. Joinson, K. Y. A. McKenna, T. Postmes, & U. D. Reips (Eds.), *The Oxford handbook of Internet psychology* (pp. 303–329). Oxford, UK: Oxford University Press.

Barendregt, B. (2013). Diverse digital worlds. In H. A. Horst & D. Miller (Eds.), *Digital anthropology* (pp. 203–224). London, UK: Bloomsbury Academic.

Bargh, J. A. (2002). Beyond simple truths: The human-internet interaction. *Journal of Social Issues, 58*(1), 1–8.

Boellstorff, T. (2013). Rethinking digital anthropology. In H. A. Horst & D. Miller (Eds.), *Digital anthropology* (pp. 39–60). London, UK: Bloomsbury Academic.

Boellstorff, T. (2016). For whom the ontology turns: Theorizing the digital real. *Current Anthropology, 57*(4), 387–407.

Brignall III, T. W., & Van Valey, T. (2005). The impact of Internet communications on social interaction. *Sociological Spectrum, 25*(3), 335–348.

Budka, P., & Kremser, M. (2004). Cyberanthropology - anthropology of cyberculture. In S. Khittel, B. Plankensteiner, & M. Six-Hohenbalken (Eds.), *Contemporary issues in socio-cultural anthropology: Perspectives and research activities from Austria* (pp. 213–226). Vienna, Austria: Löcker.

Christopherson, K. M. (2007). The positive and negative implications of anonymity in Internet social interactions: "On the Internet, nobody knows you're a dog." *Computers in Human Behavior, 23*, 3038–3056.

De Saulles, M., & Horner, D. S. (2011). The portable panopticon: Morality and mobile technologies. *Journal of Information, Communication & Ethics in Society, 9*(3), 206–216.

Ellison, N. B., & boyd, d. m. (2013). Sociality through social networking sites. In W. H. Dutton (Ed.), *The Oxford handbook of Internet studies* (pp. 151–172). Oxford, UK: Oxford University Press.

Escobar, A. (1994). Welcome to cyberia: Notes on the anthropology of cyberculture. *Current Anthropology, 35*(3), 211–231.

Gackenbach, J., & von Stackelberg, H. (2007). Self online: Personality and demographic implications. In J. Gackenbach (Ed.), *Psychology and the Internet: Intrapersonal, interpersonal, and transpersonal implications* (pp. 55–73). London, UK: Academic Press.

Gruzd, A., Jacobson, J., Mai, P., & Dubois, E. (2018). *The state of social media in Canada 2017 version 1.0*. Toronto, ON: Ryerson University Social Media Lab. Retrieved from http://digital.library.ryerson.ca/islandora/object/RULA%3A6432

Hardey, M. (2002). Life beyond the screen: Embodiment and identity through the Internet. *The Sociological Review, 50*(4), 570–585.

Kirmayer, L. J., Raikhel, E., & Rahimi, S. (2013). Cultures of the Internet: Identity, community and mental health. *Transcultural Psychiatry, 50*(2), 165–191.

Miller, D., & Horst, H. A. (2013). The digital and the human: A prospectus for digital anthropology. In H. A. Horst & D. Miller (Eds.), *Digital anthropology* (pp. 3–35). London, UK: Bloomsbury Academic.

Miller, D., & Slater, D. (2000). *The Internet: An ethnographic approach*. Oxford, UK: Berg.

Slater, D. (2002). Making things real: Ethics and order on the Internet. *Theory, Culture & Society, 19*(5/6), 227–245.

Smith, A., & Anderson, M. (2018). *Social media use in 2018*. Washington, DC: Pew Internet & American Life Project. Retrieved from http://www.pewinternet.org/2018/03/01/social-media-use-in-2018/

Wilson, S. M., & Peterson, L. C. (2002). The anthropology of online communities. *Annual Review of Anthropology, 31*, 449–467.

Chapter 8

Profound conundrums
Young lives in the digital age

Kate C. Tilleczek and Valerie M. Campbell

Introduction

Writing a book about young people in the digital age is doubly complex. The first complexity is that the lives of youth are in flux as they flow with the social, political, and economic tides of late modernity. This list of tidal shifts and surges includes rapacious capitalism, the global financial crisis, the making of a redundant labour force, and the destruction of the environment. The epistemological fallacy of late modernity (Furlong & Cartmel, 2007) also suggests that individualism is intensifying through these shifts and therefore it is becoming increasingly difficult to discern the place of such social structures and relations as they influence young lives. Young lives run parallel to a second complexity, that of apprehending the digital age as it rapidly morphs across its vast global networks. It is difficult to discern the ecologies (tools and ways of living) in a digital age of swelling technological and digital developments. For example, how do we come to know and contend with the direction, tools, and structures created by Artificial Intelligence (AI), virtual reality, automation, the Internet of Things (IoT), the push of/for Big Data, and Big Tech's[1] newest social and digital media tools as they spread across the globe? Who is controlling the Internet, how, and for what purposes? The quality of young lives, the well-being of the world's largest ever cohort, and the directions of the digital age are hotly debated and contested matters. The futures of young people hang in a balance of this complexity. The jury is quite far out.

Our *Digital Media and Young Lives* project was designed to wade into these debates in a fresh manner. Our inquiry is ongoing with further analysis and future writing projects. It was purposefully an interdisciplinary investigation that engages with the humanities and social sciences to inform youth studies. As detailed in Chapter 2, the project was planned as a form of digital and youth-attuned qualitative inquiry that provided space(s) for young people and their friends to engage in deep discussion about their relationships with technology and the digital age. In the process of this conversation, young people were very eager to engage and told us that there were too few such places for them. Whilst they can connect with like-minded networks and friends on social media, it was not seen to be the best mode for sustained or meaningful discussion or debate about digital media.

Our inquiry privileged their perspectives and ideas through means of data collection and analysis that was mindful about their place in the digital age. We agree with Kelly (2017, p. 393) in that "technology, its development, deployment and uses is never neutral ... the future has to be made/colonized" (but by whom, with what purposes, and what consequences)? If, as Bauman (2004) suggests, young people are becoming redundant in this "global spread and triumph of the modernization process" (Kelly, 2017, p. 397) what then will their futures hold? Many young people spoke, with varying degrees of hope and despair, of a future with such entities as robots, virtual reality, and holograms. Amanda[2] (21) envisioned, "Oh like a touch screen in mid air and you're just tapping stuff in mid air and everything just pops up, it's like an invisible computer walking with you at all times". Christian (17) had a less rosy view "I see it [society] coming to a downfall. Technology will take over sooo many things. Put people out of jobs. Especially with car companies. It's mostly robotic now. They're replacing a lot of workers with robots". One burning question arose from this inquiry: How are youth to enact true benefit from a digital age wherein their precariousness and well-being are in question? Our wish is to forge answers *with*, *for*, and *by* young people.

We contend that the networked landscapes within which young people dwell require new vistas for research and practice. One must read across theoretical and empirical studies relating to technology, the digital age (in all idioms and iterations of *modernity*), the running commentary about Big Tech and its power/control, and the interdisciplinary study of young people. Attempting to do so has allowed for positing questions, answers, and heuristic models (for instance, see Figures 3.1 and 6.1) which point to that which we could consider as we wade alongside young people into the digital age. This chapter illustrates persistent themes which foreground tensions and conflicted feelings about the techno-world in which youth are immersed. It also summarizes how they are actively negotiating it. Throughout the book we have placed youth voices and perspectives beside those of scholars and commentators and we have been dazzled by many similarities and some differences. One narrative arc repeats across these complex landscapes; this is a matter of grave importance to the futures and well-being of young people. We turn now to the most resonant of motifs that speak to profound conundrums of young lives in the digital age.

Hearing from youth in the digital age

It is worth noting here that quotes and examples from the same youth participants reverberate in multiple chapters. This is further evidence of the complexity of digital media in young lives. It is not just one thing, it is many things, and, for some, everything. For example, one young person could speak of their online life as challenging to their mental health as well as a privacy concern and yet digital media remains an important part of their social relationships even though they are marshalling critiques of social media platforms at the same time. And all of this is espoused by the same young person in one interview. This was the sort of conversation we invited. To hear from youth and their friends about what was/was not working and how they were brokering the digital tools across the range of their ecologies in social relationships, work, well-being, school, and so forth. The honesty with which they openly grappled with the questions was a welcome revelation to our research team.

> It's good but it's bad, because we can do anything we want now with technology. We can find anything. You can order stuff from the other side of the world, you can talk to people from the other side of the world. Like it's good because it does help with communication but it also hurts communication because nobody talks face-to-face anymore, nobody goes out anymore and just says: "Oh, let's go get a coffee". No, it's always sitting on the computer, talking to somebody over the computer and there's no real emotion there. (Rose, 18)

Many of the young people with whom we spoke expressed these sorts of conflicted feelings about digital media. They all used it to varying degrees and, while factors like convenience and long-distance communication were important to them, they were also frustrated with other aspects of online life such as social relationship drama, bullying, violence, and lack of privacy. As Henry (16) stated: "it can be good for some things, it can be bad for other things". Cameron (16) agreed:

> I think it should be banned because of the bullying problem and because of the stress problem and because of the cancer problem. But I also think it shouldn't be banned because it allows you to connect with people that you haven't connected with in years. It allows you to get easy access to your family members if you need a hold of them. It allows you to do so much, but at what cost?

One of the costs Cameron discussed was the influence that social media can have on your mental health, referring to it as a "drug" and an "addiction" as well as a space where bullying was common. This was well reported in Chapter 3. Recognizing the futility of even considering banning social media, or of getting rid of issues like bullying, Cameron concluded "it's just one of those problems that can't be fixed. 'Cause there's no way that you can tell seven billion people to be nice to each other". As also is evident in Chapter 3, the deleterious impacts that digital media has on mental health was a deep concern. They spoke of addiction, social exclusion, and violence. Many recognized that while they understood on one level that they were too dependent on their devices, they also did not make much effort to change that. Molly (17) is a perfect example: throughout the interview, she participated in the conversation, even agreeing that she might be "addicted", yet her cell phone was in her hand for the entire hour and she seldom looked up from it. She admitted to being "really hypocritical – I complain so much about using my phone. I like a good conversation with my friends but at the same time I'm complaining while I'm on my phone".

This lack of control was also highlighted by Lucy (16) who talked about a "self-control app" that would block selected sites for a period of time, supposedly to remove distraction; however, she admitted, "so then I had that but the thing is … you have to choose to turn it on, you know? That's the problem!" Control was more thoroughly discussed in both Chapters 3 and 4 and as an external problem. Participants spoke of algorithms that control what they see on their social media sites. They also recognized the permanence of their data; even though they would delete items, they were never truly deleted and could "come back to haunt" them at any time. So, although they might try to craft a certain image on social media, something they believed important for employment and educational opportunities, they could not be guaranteed that a negative post or image would not pop up at some point in the future. In contrast, there were those who believed they could control their social media image by deliberately obscuring their Internet Protocol (IP) address, using anonymous servers such as Tor, or burying specific images so deeply that the average user would not find them. These were acts of negotiation and agency that were put into play constantly, changing as the apps, media, and tools changed.

> I have a lot of directors on my Facebook like people that I work for, like administrators for [community organization] and people that I wouldn't want to view me badly so I have to control what is posted. (Marley, 17)

> I'm not applying to serious jobs right now, but that time is going to come and I figure that when that time comes, I'll probably make my profile incredibly private, same with my Twitter profile, and then I'll probably maintain it a lot more than I do now. And I'll probably go through it all and I'll sweep it for anything that could detriment myself. I'll be advertising myself to future employers. (Hunter, 20)

> In terms of public relations, you have to be popular on social media, you have to put yourself out there, you have to construct this person 'cause that is what companies will look at when they hear that you are the public relations person--they are going to look at how you interact with the world personally … your future potential employer, will look at that and say "hey, this looks like a nice person" and hire you based on that. (Gavin, 19)

Across chapters, we report that participants acknowledge the importance of digital media for communication and social relationships. When talking about well-being (Chapter 3), they referred to having a voice, which built confidence, learning about world events, and sharing ideas. And, even though concerned about online surveillance (Chapter 4), they used digital media to stay in touch with people. The students who participated in the "no phone" experiment (Chapter 5) generally felt good about living without their phones for a time, but they did miss the ability to connect with others. As a form of digital capital (Chapter 6), digital tools provided networking and job-seeking opportunities while also becoming a source of critique and concern. Digital media was also lauded as a way to make friends and stay in touch with people who were geographically distant but also as a problematic mediation that breaks down important human connections (Chapter 7).

Despite their concerns and the negative affect digital media has on their lives, the young people in this project appeared to accept the necessity of having some form of technology-mediated communication. The tools are nifty, neat, and mandatory. But they create other problems of human connection, overconsumption, and inequality which are insoluble through digital tools themselves. Participants were thoughtful and insightful about how digital media affects their lives as was evident in some of the dyadic interviews. Easton (17) and Carolyn (17) even discussed the environmental impact of discarded electronics that were shipped to China:

EASTON: It's so far away and why would it matter to us? Empathy doesn't extend over distance.

CAROLYN: Yeah, that's true. Empathy is something that happens when it's relating to you. Like, if it was your friend who was working there, you'd be like, "Oh my God, we have to stop this." But since it's some little child that you've never met in your life ... You care, but you don't care, at the same time.

In a discussion about online anonymity, Charles (19) stated, "I feel like people shouldn't let them [anonymous comments] get to them as much as people do. I know words hurt, but they should just realize that they're trolls and not listen to their comments". Bailey (21): "Don't feed the trolls".

Some, such as David in Chapter 4, appeared resigned to the role digital media plays in life. They acknowledged their inability to function in today's economy without some form of online connection, even if it was limited to checking email at the local library. There were others who more actively resisted technology's lure, either by deleting accounts or by carefully crafting their online presence to show only the portion of their lives they wanted to make public – such as their professional activities. Elijah (Chapter 3) believed that part of the problem was that young people were never provided proper education or basic understanding about the Internet and digital age; they were just expected to jump in and use it. The need for education was echoed by Willow (18) "at the same time as you're starting sex-ed with kids, I think we should have like a digital media education, and I really wanna see that starting in elementary school".

Willow's comment expressed a concern for the younger generation that is a fascinating commentary on the digital age and identified by the majority of participants. These young people worried about younger siblings, nieces and nephews, cousins, or just the *next generation* in general. Summer (17) saw younger children missing out on "childhood experiences with like, friends. They're not going out and playing, like making memories". Others felt that physical health would diminish with increased screen time, either through lack of activity or through the development of poor eating habits. This concern for future generations was particularly illuminating as it emerged organically in the interviews. When we questioned these youth as to why they were so concerned for the "kids" but less so for themselves, they pondered deeply and admitted that they hoped to be able to control digital technologies better as they grew older. They also agreed that the tools and ways of living are changing very rapidly such that the children in their lives were living a very different type of childhood from that which they had lived.

> I think, I feel like I'm lucky because I did have a childhood where I didn't just sit in front of a TV. I didn't get a phone until I was 13 or something, just to call my parents ... but now kids are growing up, they don't have to go find their friends or like call their house phone and all this ... I just think it'll disconnect people from not only other people but the world, just no sense of - I just feel bad for all the kids now. (Mckenzie, 18)

Where now?

Our project opened a space for young people to think and talk about their relationship with digital media and the digital age. The data are rich and nuanced, providing clear evidence that this is a matter of great interest and importance for today's youth. It is not an easy or simple relationship and understanding can be contradictory. Perhaps it is best summed up by Willow (18):

> for me, and for people I know, and people I talk to, digital media, social media, is a very love/hate relationship. And, I'll never figure it out, honestly. It can be abusive. It can be a loving relationship. And, just that, I have no idea what to expect from it.

We hope that this book offers some direction for the study and practices of young lives in the digital age. This project enlivens the field of youth studies by bringing together a new range of people, scholars, and commentators on the matter. But the data we collected are not clear-cut. Youth benefit from the digital tools and ecologies until they don't. And the lines between opportunity and problem remain fuzzy at best. For example, we point in Chapter 3 to a dearth of understanding digital ecologies and tools as they influence the well-being of youth. Scholars, NGOs, educators, and health practitioners are just beginning to attempt to measure this aspect of youth well-being. While we fuss about measurement and dissemination of data, youth are left largely to their own devices (no pun intended) to sort through and negotiate the digital age. And, there is plenty of money being made while they are left hanging. The CEO of Amazon (Jeff Bezos) has now become history's wealthiest person at $150 billion (USD) net worth and by bringing together technology and consumer capital under one corporation (English, 2018). The *New York Times* reports that Facebook had a 63% increase in profit and a 49% jump in revenue ($12 million) in the last quarter, and this followed the "worst crisis in its 14-year history" (Frenkel & Roose, 2018, para. 2), in which their reputation suffered following their role in the interference with US elections, the Cambridge Analytica scandal, moves towards further regulation, and a poorly landed apology by its CEO (Mark Zuckerberg).

> The Cambridge Analytica scandal engulfed Facebook, sending the company's stock price plunging and setting in motion the worst crisis in the company's history. Cambridge executives had long bragged about deploying powerful "psychographic" voter profiles to manipulate voters. Now Facebook was forced to acknowledge that Cambridge had used voters' own Facebook data to do it. The damage was not only legal and political — Facebook faced lawsuits and new inquiries by regulators in Brussels, London and Washington — but also reputational. Silicon Valley's public image had survived the Snowden revelations. But tech companies, already implicated in the spread of "fake news" and Russian interference in the 2016 election, were no longer the good guys. (Confessore, 2018)

However, it seems that even a disastrous year for Facebook's public relations could not harm its profits. With this much money to be gained from the digital age and this much political power in the hands of technology companies and their political lobbies, how are youth to stay in control of their futures, become informed and knowledgeable about the world in which they live, and remain selective about that which they will/will not partake? How will their agency, activism, and negotiation be understood and supported?

In short, there must be an educational revolution for youth in the digital age. It must place the well-being of youth over corporate profit and refuse to be led by corporate and Big-Tech's ambitions. The details of such a revolution will be presented in a future book and informed by the perspectives of young people as we further mine data and scholarship. Suffice it to state that directions will be of the order that Tilleczek (2011, 2012, 2016; Tilleczek & Ferguson, 2013) suggests; that it not be technocratic education for technicians, not standard education *for* young people, not private education for capital gain, but rather a democratic education *with*, *for*, and *by* young people in the digital age and towards global good. In this goal, we share the waning optimism of Berners-Lee that we are not too late. As the inventor of the World Wide Web in 1990 and the recent recipient of the Turing Award in computer science, Berners-Lee today speaks openly about how his plan for web technology "that allows anyone to share information, access opportunities and collaborate across geographical boundaries" (Solon, 2017, para. 2) has been called into real question. The digital age, its tools, and its ecologies have been overshadowed by control of the Internet and digital media by those who are unconcerned with the good of humanity and democracy.

> I'm still an optimist, but an optimist standing at the top of the hill with a nasty storm blowing in my face, hanging on to a fence. We have to grit our teeth and hang on to the fence and not take it for granted that the web will lead us to wonderful things. (Solon, 2017, para. 3–4)

Berners-Lee's cautious optimism was shared by many of our young participants. We give the final word to Piper (17): "It's good that technology is advancing fast because then maybe it will help some for a good cause. But also then there's the downside of ... how do you control it?"

Note

1 Facebook, Google, Apple, Amazon, and Microsoft (Lotz, 2018).
2 All participants were given a pseudonym.

References

Bauman, Z. (2004). *Wasted lives: Modernity and its outcasts*. Cambridge, UK: Polity.
Confessore, N. (2018, Aug 14). The unlikely activists who took on Silicon Valley – and won. *New York Times*. Retrieved from https://www.nytimes.com/2018/08/14/magazine/facebook-google-privacy-data.html
English, C. (2018, Jul 17). Jeff Bezos is the world's richest person – and it's not even close. *New York Times*. Retrieved from https://nypost.com/2018/07/17/jeff-bezos-is-the-worlds-richest-person-and-its-not-even-close/
Frenkel, S., & Roose, K. (2018, Apr 25). Facebook's privacy scandal appears to have little effect on its bottom line. *New York Times*. Retrieved from https://www.nytimes.com/2018/04/25/technology/facebook-privacy-earnings.html
Furlong, A., & Cartmel, F. (2007). *Young people and social change: New perspectives*. Buckingham, UK: Open University Press.
Kelly, P. (2017). Young people and the coming of the third industrial revolution. In A. Furlong (Ed.), *Routledge handbook of youth and young adulthood* (2nd ed., pp. 391–399). Abingdon, UK: Routledge.

Lotz, A. (2018, Mar 23). 'Big Tech' isn't one big monopoly – it's 5 companies all in different businesses. *The Conversation*. Retrieved from https://theconversation. com/big-tech-isnt-one-big-monopoly-its-5-companies-all-in-different-businesses-92791

Solon, O. (2017, Nov 16). Tim Berners-Lee on the future of the web: 'The system is failing'. *The Guardian*. Retrieved from https://www.theguardian.com/technology/2017/nov/15/tim-berners-lee-world-wide-web-net-neutrality

Tilleczek, K. (2011). *Approaching youth studies: being, becoming, and belonging.* Toronto, ON: Oxford University Press.

Tilleczek, K. (2012). Policy activism with and for youth in transition through public education. *Journal of Educational Administration and History, 44*(3), 253–267. doi:10.1080/00220620.2012.683393

Tilleczek, K. (2016). Voices from the margins. *Education Canada, 56*(4), 34–39.

Tilleczek, K., & Ferguson, B. (Eds). (2013). *Youth, education and marginality: local and global expressions.* Waterloo, ON: Wilfred Laurier University Press.

Index

Note: Boldface page numbers refer to tables; italic page numbers refer to figures and page numbers followed by "n" denote endnotes.

3 20